The GOD Gene

The GOD Gene

How Faith Is Hardwired into Our Genes

Dean Hamer

Doubleday

New York London Toronto Sydney Auckland

PUBLISHED BY DOUBLEDAY
a division of Random House, Inc.

DOUBLEDAY and the portrayal of an anchor with a dolphin are
registered trademarks of Random House, Inc.

Book design by Chris Welch

Library of Congress Cataloging-in-Publication Data
Hamer, Dean
The God gene / Dean Hamer.—1st ed.
p. cm.
Includes bibliographical references and index.
1. Psychology, Religious. 2. Genetic psychology. I. Title.
BL53.H285 2004
200'.1'9—dc22 2004047808
ISBN 0-385-50058-0

PRINTED IN THE UNITED STATES OF AMERICA

October 2004
First Edition
1 3 5 7 9 10 8 6 4 2

To Donald and David,
my spiritual guides

Acknowledgments

Religion and spirituality are topics far removed from my professional expertise in molecular genetics. Without the help and encouragement of friends and colleagues from many disciplines, I could never have written this book.

I owe special thanks to Ronald Green, professor of religion and chair of the Ethics Institute at Dartmouth University, who early on helped me focus on the personal aspects of religion. Ron steered me to the right questions, even though he still fundamentally disagrees with my answers. Gerald Edelman, the Nobel laureate biologist who now directs the San Diego Neuroscience Institute, is another skeptic who nevertheless provided me with deep insights into the brain and consciousness.

Robert Cloninger, a physician-scientist who is professor of psychiatry, genetics, and psychology at Washington University School of Medicine in St. Louis, played an important role in convincing me that spirituality is a measurable quantity. His self-transcendence scale is central to the book, and I have always profited by our interactions.

Lindon Eaves, who is both a professor of human genetics and psychiatry at Virginia Commonwealth University and an Anglican priest, provided much insight and knowledge about how spiritual and religious values are transmitted. I thank Susan Blackmore, a senior lecturer in psychology at the University of the West of England, Bristol, for being so unselfish in communicating her insights into memes. Lee Dugatkin, associate professor in the Department of Biology at University of Louisville, Kentucky, and author of several books about animal behavior, greatly impacted my thinking about evolution and group selection.

Both Andrew Newberg, director of nuclear medicine at the Hospital of the University of Pennsylvania, and V. S. Ramachandran, who is director of the Center for Brain and Cognition at the University of California, San Diego, and adjunct professor of biology at the Salk Institute, provided insight into the ingenious neurobiological methods they have developed to probe the role of the brain in spirituality and belief. On the molecular side, George Uhl at the National Institute on Drug Abuse was kind enough to share with me his vast knowledge of monoamine transporters and their genes, including VMAT2.

The chapter on the DNA of the Jews could not have been written without the help of David Goldman and Neil Bradman in London, England, and Karl Skorocki in Haifa, Israel, who were kind enough to spend time with me explaining both the science and the fascinating personal stories behind their research. Also in Israel, Bob Belmaker, Jon Benjamin, and Dick Ebstein provided both sage advice on psychiatric genetics and a guided tour of the multicultural synagogues of Beersheba.

I learned much about witchcraft and the religious construction of gender from my cousin Wendy Griffin, professor of women's studies at California State University, Long Beach. Through

Wendy, I met the formidable high priestess and teacher of the Wiccan way, Vivianne Crowley, who lectures and writes about paganism in England.

I had the great fortune to meet James Austen, author of *Zen and the Brain*, at a conference in Kyoto, where he provided sage advice on both the neurobiology and the practice of Zen Buddhism. It was during the same trip that I visited the Hosenji Zen Center, where Tenkai and several monks were kind enough to discuss their own unique insights into Zen.

Julie Castiglia, my agent, and Roger Scholl, my editor, were instrumental in convincing me that people might actually read a book that mixed the seemingly incompatible topics of religion and science. I thank them for their guidance, patience, and formidable editing skills.

I owe thanks to several organizations and institutions as well as individuals. The Templeton Foundation is a terrific source for information about religion and science through their Web site, newsletters, and sponsored lecture series. The Zygon Center for Religion and Science publishes *Zygon*, the premier academic journal on religion and science, while Science and Spirit Resources publishes the accessible periodical *Science and Spirit*.

For the past twenty-eight years, the National Institutes of Health has provided me with a superb research environment and many talented colleagues. I would like to emphasize, however, that this book was written strictly as an outside activity with no financial support or institutional encouragement from the United States government. The ideas expressed here are solely my own.

My greatest inspiration was William James, whose *Varieties of Religious Experience* has provoked my thinking about the role of biology in culture ever since I first read it for a psychology course as an undergraduate. His belief that there is a fundamental divide between private and public religion is central to my thinking. Although James lived long before genes and DNA were discovered, I like to think that he would fundamentally agree with the thesis of *The God Gene*.

Lastly, I thank my partner, Joe Wilson, and my spiritual guides, Donald and David, for standing by me through thick and thin.

Contents

The GOD Gene

One

Spiritual Instinct

Instinct leads, logic does but follow.
—*William James*

The first thing I noticed about Tenkai was his smile. It was serene, content, knowing but not smug. It was the smile of a person at peace with himself and the world around him. The smile of someone who had seen much but could still be surprised. It was a spiritual smile.

The second thing I noticed was that even though Tenkai spoke fluent Japanese, wore traditional Japanese garb, and was living in a Japanese monastery, he was clearly not Asian. His blue eyes and light brown hair were the giveaway. He, in fact, was born and raised in Hamburg, Germany, as Michael Hoffman.

I met Tenkai at the Hosenji Zen Center, which is located in a small Japanese village about an hour's train ride west of Kyoto. The Center provides a venue for people from different countries and religious traditions to learn about Zen Buddhism and practice its system of meditation, known as *zazen,* in which one sits in the lotus position with half-closed eyes and focuses on breathing. The idea is to empty the mind of all thoughts.

I participated in the Center's daily activities: waking to a gong at 6 A.M., an hour of zazen sitting in a traditional tatami mat room overlooking a waterfall, eating a breakfast of rice gruel and pickled vegetables in silence, several hours of weeding the vegetable garden or sweeping the stone paths, sutra chanting, supper of more rice and vegetables, and a final two hours of meditation in a log cabin overlooking a temple. It was a peaceful life. But I was not at the Center to ripen my intuitive faculties or to experiment with monastic existence. I was there attempting to understand whether or not there is a biological basis of spirituality.

Until the age of twenty-four, Tenkai led an ordinary life as a high school teacher. Following a breakup with his girlfriend, however, he began to ask himself the life questions at the heart of our need to believe in something larger than ourselves. Why was he here? What is the purpose of life? Why is there so much suffering? Soon thereafter he quit his job, got on his bicycle, and started pedaling east. He didn't stop until he reached the shores of the East China Sea.

Along the way, Tenkai experimented with every spiritual system and mystical tradition he encountered. In Austria, he studied the principles of anthroposophy, which claims to be "a path of knowledge which leads the spiritual in the human being to the spiritual in the universe." In an Indian ashram, he practiced a type of meditation in which bouts of exuberant dancing and singing were alternated with periods of complete motionlessness. He prayed for twelve hours a day at a monastery in Nepal, and in China he practiced the graceful motions of Tai Chi.

At times he fasted and abstained from sex, while at other times he mixed alcohol, drugs, and women with abandon. There were times when he sat quietly for hours; at other times he jumped about and grunted like a gorilla. But no matter what he tried, no matter which spiritual leader he followed, Tenkai felt that something was missing. It wasn't until he abandoned his bicycle and flew to Japan that he found what he was looking for: Zen Buddhism.

Zen is based on the premise that every human being is capable of enlightenment. All that is needed is to see through to one's true nature through meditation. The emphasis is on living in the present with no regret for the past or fear of the future. The motto on the Hosenji Zen Center's brochure, for example, is "Your future is here now."

Zen is unique among religions in that it is virtually devoid of theology, scripture, or ritual. There are no gods or devils, there is no heaven or hell. Although Zen does have priests, they do not have any special claim to holiness.

When I met Tenkai, he had just completed his apprenticeship at the traditional monastery of Empuku-ji and started residency at the Hosenji Center. The monastic life seemed to suit him well. Slowly, he told me, things began to change for him. It wasn't so much that the monastic life changed the way he felt—he still had emotional ups and downs—or the way he thought. Zen isn't about cognition, he explained. Instead it changed the way he perceived things. He felt he was more integrated with the universe, his sense of reality more focused. What changed, he says, was his sense of the way things are—his consciousness.

Sometimes this new sensibility came upon him when he was practicing zazen. Other times it occurred when he was performing hard physical labor. Asked what it felt like, he told me, "I can't explain it in words, but once you have the feeling you'll understand. It's like your mind just isn't there."

When I admitted to Tenkai that I spent much of my meditation time fretting about routine matters, he gave me a spiritual smile. "It's a strange and mystical thing, this feeling. But don't worry," he said. "Everybody gets it at one time or another."

A Human Universal

Although Tenkai may be unusual compared to the average person in the strength and tenacity of his metaphysical yearnings, he is by no means unique. Most people, psychologists and theologians would argue, have some capacity for spirituality. It is among the most ubiquitous and powerful forces in human life. It has been evident throughout recorded history in every civilization and culture, in every nook and cranny of the globe. For many people, it is the main focus of their lives.

Homo sapiens have had spiritual beliefs since the dawn of our species. More than 30,000 years ago, our ancestors in what today is Europe painted the walls of their caves with images of strange chimeras with human bodies and animal heads, representing, anthropologists feel, sorcerers or priests. These early humans buried their dead with elaborate preparations for the afterlife. Sometimes they equipped them with food and supplies for their journey; other times they cut off their hands and heads, perhaps to prevent the return of enemies. These are the actions of believers.

Our spiritual convictions remain just as strong today. Surveys show that more than 95 percent of Americans believe in God, while 90 percent meditate or pray, 82 percent say that God performs miracles, and more than 70 percent believe in life after death.

Our faith is not unique. Even in China and the former Soviet Union—where powerful governments used every possible form of persuasion to replace God with Communism—more than half of the people retained their spiritual beliefs. Meanwhile, the forces of fundamentalism—whether Christian, Jewish, or Muslim—are sweeping across the globe from South America to the Middle East to Africa.

Our spiritual beliefs are not necessarily reflected in terms of church attendance, however. In fact, church attendance has been declining in the United States ever since the 1950s. More and more people feel that churches place too much emphasis on organization and not enough on spirituality. As noted in one Gallup poll, "believing is becoming increasingly divorced from belonging."

Church attendance in Europe has declined even more dramatically. In England and Holland, only 5 percent of the population attends religious services on a regular basis. Even in Italy, often considered a bastion of traditional Catholic values, the majority of people disagree with the Pope on religious issues such as divorce and abortion.

The discrepancy between flagging attendance in formal religious settings and the high percentage of people who believe in God can be explained in large measure by the fact that spirituality is distinct from the precepts of any particular religion. More than three-quarters of people surveyed feel that God can reveal himself through many different paths. There is a growing sentiment that it doesn't matter what church you go to because "one is as good as the other."

Despite formal religious adherence or attendance, a survey taken in the 1990s found that 53 percent of Americans have had a "moment of sudden religious awakening or insight." In our own research at the National Institutes of Health, more than one-third of the essentially random collection of people we surveyed reported personal experiences in which they felt in contact with "a divine and wonderful spiritual power." A similar percentage of people said that they have, at least once or twice, felt "very close to a powerful, spiritual force that seemed to lift you out of yourself." Although such experiences were once regarded as signs of incipient psychopathology, recent research shows that they actually are associated with better adjustment and psychological health in most people.

While someone like Tenkai may be at the high end of the human continuum in terms of his interest in spirituality, he is by no means unique. He stands out in the degree of his spirituality rather than in the fact that he believes in something larger than himself. He just has more than the usual degree of what in fact is a common human trait.

The God Gene Theory

Why is spirituality such a powerful and universal force? Why do so many people believe in things they cannot see, smell, taste, hear, or touch? Why do people from all walks of life, around the globe, regardless of their religious backgrounds or the particular god they worship, value spirituality as much as, or more than, pleasure, power, or wealth?

I argue that the answer is, at least in part, hardwired into our genes. Spirituality is one of our basic human inheritances. It is, in fact, an instinct.

At first, "instinct" may seem to be a peculiar word to pair with spirituality. We usually think of instincts as automatic, unconscious reactions or behaviors that are performed without thought or training. Birds know to fly south for the winter by instinct. Blinking your eyes when someone takes a swing at you is an instinct. A newborn baby learns to suckle at her mother's breast by instinct, not because she has been taught. It is instinctual to become aroused when presented with a sexual stimulus. By contrast, spiritual *behaviors* such as meditating in the lotus position or taking communion are neither automatic nor unconscious. They are highly deliberate and culturally learned activities.

I do not contend that spirituality is a simple instinct like blinking or nursing. But I do argue that it is a complex amalgamation in which certain genetically hardwired, biological patterns of response and states of consciousness are interwoven with social, cultural, and historical threads. It is this interdigitation of biology and experience that makes spirituality such a durable part of the fabric of life—a rich tapestry in which nature is the warp and nurture is the woof.

The idea of complex characteristics that are influenced by both genes and environment is not unique to spirituality. There are many well-known examples of the interplay between nature and nurture, even among animals. Consider the song of the white-crowned sparrow. The male sparrow's song, which he begins to sing at about seven or eight months of age, consists of a long, low whistle followed by a series of trills. That the song is at least partly innate can be seen by its species specificity: All male white-crowned sparrows sing basically the same song, even when they are separated from their parents after two weeks and raised in complete isolation from any other birds. Moreover, there is a dedicated brain circuit for song, consisting of six interconnected brain regions that control the vibrations of the vocal membranes in the throat. (Three of the regions are sensitive to sex hormones, which explains why only males produce the characteristic mating song.) The reason a sparrow sings the song of a sparrow, and not of a robin or a lark, is that it has the genes and brain of a sparrow, not that it was raised by sparrows.

But experiments show that the song is also partly learned. Sparrows from different areas sing slightly different varieties of the song, or "dialects," that differ in the exact number and placement of the trill notes. When sparrows are left with their parents for the first three months of life, then raised in isolation or with foster parents from another region, they still sing the parental dialect when they begin to vocalize a few months later. By contrast, when the young birds are taken from the nest after only a few weeks, then raised hearing tape recordings of a different dialect, they sing the foreign song.

An even more profound effect of the environment has been shown in studies in which the chicks have been deafened at an early age. These poor animals never produce anything more than a few unconnected notes, no matter how hard they try. They need to hear themselves to sing properly.

Such experiments show that while the basic species-specific skeleton of the song is hardwired in the genes, it requires the environmental clue of being able to hear its own voice to be triggered. And while the precise details of the song are culturally transmitted during a critical period early in life, they require the right biological machinery and genetic code to be actualized. It's part of the remarkable yin and yang of nature and nurture, especially in a species that most people would never even associate with having a culture.

In *The God Gene*, I propose that spirituality has a biological mechanism akin to birdsong, albeit a far more complex and nuanced one: that we have a genetic predisposition for spiritual belief that is expressed in response to, and shaped by, personal experience and the cultural environment. These genes, I argue, act by influencing the brain's capability for various types and forms of consciousness, which become the basis for spiritual experiences.

The term "God gene" is, in fact, a gross oversimplification of the theory. There are probably many different genes involved, rather than just one. And environmental influences are just as important as genetics. Finally, spirituality, in its broader meaning, is about much more than belief in a particular God. Some of the most spiritual people I've interviewed and discuss, such as Tenkai, don't believe in a deity at all. Nevertheless, I felt it was a useful abbreviation of the overall concept.

The Fivefold Way

Proving there is a genetic component to spirituality is no simple task, and probably no single line of evidence or observation will be completely convincing. It's not like hair or eye color, which are passed from one generation to the next in an obvious way. Therefore, I use several different lines of reasoning and types of data in the book to show the instinctual side of spirituality. Some of them are based in traditional approaches to religion such as psychology and anthropology. The main emphasis, however, is on powerful new research methods that have been developed in molecular genetics and neurobiology. The proof, in other words, depends on the whole pudding—the entire array of evidence—not just one ingredient. Let me summarize the five essential arguments I intend to present:

Measurement. The first task for any scientist attempting to link genetics to spirituality is to show that spirituality can be defined and quantitated. This is essential for any scientific analysis, regardless of the topic. Scientists measure things. If we can't measure it, we can't test a hypothesis about it, and if we can't test a hypothesis, it can't be proved.

Measuring spirituality is particularly difficult because it encompasses so many different types of feelings, beliefs, and experiences. Fortunately, a number of psychologists have tackled the problem using sophisticated statistical methods of psychological measurement. In the book, I use a scale called "self-transcendence" developed by Robert Cloninger, an innovative thinker who studies the biological and social origins of personality.

Self-transcendence provides a numerical measure of people's capacity to reach out beyond themselves—to see everything in the world as part of one great totality. If I were to describe it in a single word, it might be "at-one-ness."

Although self-transcendence might seem a bit "flaky" to some readers, it successfully passes the tests for a solid psychological trait. It basically is a yardstick for what is often referred to in the West as *faith,* or in the East as the search for *enlightenment.*

One of my biggest challenges in *The God Gene* is to attempt to separate spirituality from religion. This is difficult because religion is founded on spiritual beliefs. Conversely, spiritual beliefs usually are expressed using the language and rituals of religion. Nevertheless, the self-transcendence scale tries to separate one's spirituality from one's particular religious beliefs by eschewing questions about belief in a particular God, frequency of prayer, or other orthodox religious doctrines or practices. Even individuals who dislike all forms of organized religion may have a strong spiritual capacity and score high on the self-transcendence scale.

Heritability. The next task for the scientist exploring the link between genetics and spirituality is to determine whether spirituality is inherited, and if so, to what extent. This can be tackled by using twin studies—comparing identical twins, who have the exact same genes, to fraternal twins, who are only as genetically similar as ordinary siblings. One can determine a factor's *heritability* by comparing resemblances and differences between different types of twins, and between twins and unrelated people.

Scientists have used twin studies to show that spirituality, as measured by the self-transcendence scale, is significantly heritable. The extent of genetic influence is similar to that for many personality traits, and even greater than it is for some physical characteristics. In other words, there is a strong genetic link.

Twin studies also can be used to study the role of the environment on a trait or behavior. Not surprisingly, upbringing plays an important role in spirituality. But remarkably, what counts most is not the specific shared cultural environment, such as Sunday school, sermons, or parenting. What is important are the unique life events each person experiences on their own.

Identifying a Specific Gene. While twin studies suggest that spirituality is partially inherited, they say nothing about which genes are involved or how they work. That's the job of molecular biology. One major new finding revealed in *The God Gene* is our discovery of a specific individual gene associated with the self-transcendence scale of spirituality. This "God gene" codes for a monoamine transporter—a protein that controls the amount of crucial brain signaling chemicals. Interestingly, these same brain chemicals can be triggered by certain drugs that can bring about mystical-like experiences.

The specific gene I have identified is by no means the entire story behind spirituality. It plays only a small, if key, role; many other genes and environmental factors also are involved. Nevertheless, the gene is important because it points out the mechanism by which spirituality is manifested in the brain.

Brain Mechanism. The monoamines I identify—the brain chemicals controlled by the God gene—have many different functions in the brain. They appear to influence spirituality by altering consciousness, which can be broadly defined as our sense of reality—our awareness of ourselves and the universe around us, including our thoughts, memories, and perceptions.

The intimate relationship between spirituality and consciousness becomes most conspicuous through mystical experiences, such as Saul's conversion on the road to Damascus, which are invariably accompanied by major alterations in perceptions. But the relationship between spirituality and consciousness also is evident in more subtle forms, such as self-transcendence, in which consciousness is altered by a blurring of the normal distinction between self and other.

Monoamines such as serotonin and dopamine are important players in consciousness. According to a theory developed by Gerald Edelman, the key role of monoamines with regard to consciousness is to link objects and experiences with emotions and values. Evidence supporting the importance of monoamines in affecting consciousness may be seen with the help of sophisticated brain-scanning techniques and by analyzing the actions of various types of drugs that block or enhance these brain chemicals, as well as in studies of individuals with brain lesions such as temporal lobe epilepsy.

Selective Advantage. Darwin's theory of evolution and competitive advantage applies as much to complex human behavioral characteristics such as spirituality as it does to beak shape in birds or hunting ability in lions. What are the selective advantages of having God genes? Are they simply a side effect of the evolution of the mind, or do they offer us a more direct evolutionary advantage?

I argue that one of the important roles that God genes play in natural selection is to provide human beings with an innate sense of optimism. At the psychological level, optimism is the will to keep on living and procreating, despite the fact that death is ultimately inevitable. At the physical level, studies show that optimism seems to promote better health and quicker recovery from disease, advantages that would help us live long enough to have and raise children and pass on our genetic heritage.

These five lines of reasoning and evidence are the heart of the book. In the last part of *The God Gene,* I turn to the broader questions posed by the interface of spirituality and biology with religion and society.

From Spirituality Genes to Religious Memes

Spirituality is an intensely personal activity. It involves private feelings, thoughts, and revelations. These are often difficult if not impossible to describe, much less to share. Yet only rarely does spirituality occur in a complete vacuum; even the most isolated ascetic must sometimes come in contact with other people. More often than not, spirituality is associated with a much more public domain of human life: religion.

Of course, religion in human society is far more than just a public manifestation of spirituality. Religious institutions act as schools, courts, sanctuaries, landholders, and counselors, as well as places of worship and prayer. What do these diverse functions have to do with the biology of spirituality?

To find out, scientists used the same experimental approaches described earlier to show that spirituality is heritable and applied them to traditional religious behaviors and attitudes. The results showed that although religiousness *does* have a genetic component, it is much weaker than that for spirituality. Religion, unlike spirituality, is transmitted primarily not by genes, but by *memes*: self-replicating units of culture, ideas that are passed on from one individual to another through writing, speech, ritual, and imitation.

While it is our genes that initially make us receptive to spirituality and faith, it is our memes that carry religion from one generation to the next and that make each religion distinct. Later, I discuss how such memes work, and the ways in which spirituality glues them together into coherent and enduring religious institutions.

One of the strongest of all religious memes is the concept that it is a sin to marry a nonbeliever. In Chapter Ten, I show the extent to which one group, the Jews, have embraced and followed this proscription. DNA fingerprinting demonstrates that Jews all around the world—from the Middle East to the Midwest, from the Jewish ghettos of Europe to the villages of sub-Saharan Africa—have preserved the same telltale pattern of DNA snippets. Despite the diaspora, the Inquisition, and the Holocaust, Jews have sustained their genetic heritage as well as their religious traditions. Remarkably, the coalescence of these telltale DNA sequences can be dated to approximately 3,000 years ago—the time of the exodus from Egypt, according to the Bible.

In the final chapter, I consider the findings of *The God Gene* in the context of human history. Spiritual beliefs have been with us since recorded history and, despite the growth of scientific inquiry in the past century, show no signs of weakening. God is not dead, to answer *Time* magazine. There is, however, a growing tendency to pit science and spirituality (or religion) against each other as if they were intrinsic enemies. They are not. As Albert Einstein famously commented, "Religion without science is blind; science without religion is lame."

While the question "Is there a God?" may be beyond science, the question "*Why* do we believe in God?"—in other words, an understanding of the mechanism through which our belief in God or a higher power works—is potentially within our understanding.

Caveats

Every good scientific treatise or book has a section on "limitations"—a listing of potential weaknesses, flaws, and ambiguities in the data and its interpretation. This is part of what makes the scientific method so powerful. Scientists do not jump to conclusions without looking at the quality of the evidence. (If only candidates for political offices listed their limitations in the same way!)

This book has three major limitations that I want to emphasize up front.

First, *this is not a complete explanation of spirituality*. Genes can account for part of the story of spirituality, but not all of it—not by a long shot. For one thing, the key empirical research is based on a single measure of spirituality, the self-transcendence scale. As I describe in the next chapter, self-transcendence is a valid, robust, and fairly general yardstick for spirituality; high scorers, like Tenkai, whom I introduced at the beginning of the book, would be considered spiritual by most people. But such a scale is by no means comprehensive. Spirituality is too multifaceted to be captured in its entirety by a single measure.

Furthermore, genes explain only about half of the variation that is seen even for this one scale. And the single gene I've identified is responsible for less than that—a small percentage of variance at best. That means that most of the observed differences are still unaccounted for. This is the reason that spirituality can never be an all-or-nothing trait. It's not like eye color, where the result is either one thing or the other. It's more graduated, like height.

My second caveat is that *behavioral genetic research can explain only individual differences, not species-specific characteristics*. What that means is that the genetic data can help explain why Tenkai is more spiritual than somebody else, but not why humans in general are spiritual (whereas most other life-forms presumably are not). Of course, the hope is that the same genes that play a role in individual differences also are key players in the overall trait. That's why I extrapolate from our finding that a monoamine transporter gene plays an important role in individual differences in self-transcendence to a more general theory about the neurobiology of consciousness and spirituality. Strictly speaking, though, species-specific traits are the domain of evolutionary psychology, a fascinating but less experimentally rigorous discipline.

My third caveat is the most important. *This is a book about why humans believe, not whether those beliefs are true.* Nonbelievers will probably argue that finding a God gene proves there is no God—that religion is nothing more than a genetic program for self-deception. Religious believers, on the other hand, can point to the existence of God genes as one more sign of the creator's ingenuity—a clever way to help us humans acknowledge and embrace his presence.

Both of these arguments mix apples with oranges, or in this case, theology with neurobiology. The one thing we know for certain about spiritual beliefs and feelings is that they are products of the brain—the firing of electrochemical currents through networks of nerve cells. Understanding how such thoughts and emotions are formed and propagated is something that science can tackle. Whether the beliefs are true or false is not. Spirituality ultimately is a matter of faith, not of genetics.

I know from experience that some readers will ignore this caution, so I'll repeat it for good measure. This book is about whether God genes exist, not about whether there is a God.

Two

Self-Transcendence

Feeling is the deeper source of religion.

—*William James*

As I began my research into spirituality, the student center at George Mason University in Reston, Virginia, became a frequent stop. Not that it's a particularly reverent place—on one typical spring afternoon it was full of students chatting, studying, checking e-mail, and eating fast food. There was a booth selling tickets to a rock concert, another with information about the ROTC. But my interest was focused on a side lounge festooned with advertisements for research volunteers. My own advertisement was among them; it read "Earn 40 Dollars!" (College students will do a lot for a little cash.) The fine print explained that my colleagues and I were looking for pairs of siblings to participate in a genetics study; participants would need to give us a blood sample as well as answer some questions.

Inside the lounge, a dozen or so young men and women dressed in various combinations of baseball caps, shorts, jeans, and sandals were sitting at tables. They were filling out a questionnaire called the Temperament and Character Inventory, or TCI for short, which is a 240-question true-false quiz that assesses seven dimensions of personality. My colleagues and I were collecting the information for a study of the genetics of cigarette smoking, which was my chief research project at the National Cancer Institute. But the TCI also happens to include a scale to measure self-transcendence. That's where the spirituality part came in.

Self-transcendence is a term used to describe spiritual feelings that are independent of traditional religiousness. It is not based on belief in any particular God, frequency of prayer, or other orthodox religious doctrines or practices. Instead, it gets to the heart of spiritual belief: the nature of the universe and our place in it. Self-transcendent individuals tend to see everything, including themselves, as part of one great totality. They have a strong sense of "at-one-ness"—of the connections between people, places, and things. Non-self-transcendent people, on the other hand, tend to have a more self-centered viewpoint. They focus on differences and discrepancies between people, places, and things, rather than similarities and interrelationships.

Self-transcendence would become a key concept in my search for the God gene. It's the yardstick scientists use to gauge the intensity of individuals' spiritual feelings. At first it might seem simplistic to measure something as complex as human spirituality by an instrument as quaint as a paper-and-pencil self-questionnaire. Wouldn't it be more accurate to look at actual behavior (what people really do) rather than self-perceptions (the way they say they feel)?

If our intent had been to measure religiousness rather than spirituality, that clearly would have been an option. We might have explained how often people attended religious services, for example, or whether they took their children to Sunday school. But as William James pointed out in *The Varieties of Religious Experience*, feelings are what count when it comes down to the sort of private religiousness that we now call spirituality. And since we don't yet have any mechanical device that can read a person's feelings, asking questions is still the best we can do. The critical issue is which questions to ask. That's where psychology comes in.

The Atheist's Discovery

One of the first modern psychologists to tackle the problem of measuring spirituality separate from religion was an avowed atheist, Abraham Maslow.

Maslow was the founder of humanist psychology, a school of thought that became popular during the socially turbulent 1960s. Unlike most of his colleagues, Maslow was more interested in the good than the bad side of our mental life. He focused on human potential instead of psychopathology or mental disorders. To do so, he studied people he believed had actually realized their psychological potential. He called such people self-actualizers.

Maslow's criteria were strict. Thomas Jefferson, Eleanor Roosevelt, and Albert Schweitzer met the grade. Benjamin Franklin and Walt Whitman did not; they were considered to be only potential or possible cases. Out of 3,000 college students that Maslow surveyed, just one met his definition of a true self-actualizer.

Maslow noted many similarities between the self-actualizers he identified. They were spontaneous. They looked at even ordinary things afresh, as if for the first time, and had the ability to see things as they really are, regardless of physical details or other people's opinions. They were highly ethical but not always conventional. Self-actualizers were also empathetic. They identified and sympathized with people from all walks of life. Many of them extended their compassion to other living things and to nature generally.

As Maslow studied more and more self-actualizers, he realized that they shared one other key feature. They underwent periodic spiritual experiences.

This likely came as a surprise to Maslow, who took pride in being an atheist and compared belief in God to "the childish looking for a big Daddy in the sky." Perhaps that is why he gave these events a new name: peak experiences. But call them what you will, the descriptions are strikingly similar to those of spiritual revelations reported by both Western religious figures and Eastern meditators. For example, Maslow quotes one of his subjects as saying:

> I could see that I belonged in the universe and I could see where I belonged in it; I could see how important I was and yet how unimportant and small I was, so at the same time that it made me humble, it made me feel important.

The key feature of peak experiences is a sense of wholeness and unity with the universe, with everything and everyone. There is an effortless letting go of the ego, a going beyond the self. Spontaneity and creativity are enhanced, personal and physical needs are forgotten. Things are seen as they *really* are, not as they serve the viewer's needs. Maslow called this way of thinking and feeling "being-cognition," to distinguish it from the "deficiency-cognition" that characterizes ordinary consciousness.

Maslow maintained that organized religion hindered rather than helped people become self-actualizers. Given his atheism, however, this may have just been prejudice. Can peak experiences and being-cognition really be separated from more conventional religiousness?

To find out, psychologist E. L. Shostrom developed a new questionnaire—the "Personal Orientation Inventory: An Inventory for the Measurement of Self-Actualization"—and started giving it to subjects. As predicted by Maslow, there was an inverse relationship between scores on this inventory and orthodox religiousness. In one study, for example, college students who had received religious training at parochial high schools scored lower than graduates of nonreligious schools. Other investigations showed that individuals who scored high on the self-actualization inventory were *less* likely to attend church, but more likely to have mystical experiences, than were low scorers. Catholic priests scored lower than laypeople, although transcendental meditators scored higher than nonmeditators.

These empirical studies underscored the fact that being-cognition—Maslow's version of spirituality, though he never would have called it that himself—is fundamentally different from orthodox religiousness.

Unfortunately, however, the "Personal Orientation Inventory" turned out to be a poor psychological instrument; the subscales didn't always hang together, the association with mystical experiences was not reliably replicated, and scores went down instead of up with age and experience. A better measuring stick was needed. It took another three decades to invent it.

A Spiritual Yardstick

Robert Cloninger, a psychiatrist at Washington University Medical School in St. Louis, is perhaps the first modern-day behavioral scientist bold enough—or foolhardy enough—to have tried to quantitate spirituality. He is the inventor of the self-transcendence scale, which grew out of a system of personality classification called the *biosocial model,* and is measured as part of the Temperament and Character Inventory.

Cloninger, a tall and thin man who walks with a slight hunch, has the distracted, absent-minded air of a beloved professor, an image enhanced by his southern manners and soft-spoken, formal style of conversation. His harmless appearance has fooled plenty of scientists who mistook it for a lack of intellectual prowess.

In truth, Cloninger is a stubborn, tough advocate for his ideas and a formidable debater. Woe to the colleague who comes unprepared; his expansive mind is filled with ready facts, references, case histories, theories, and statistics. They are frequently needed, for Cloninger's ideas about the importance of spirituality are not universally accepted by his peers. In fact, most modern psychiatrists and psychologists would just as soon not explore or discuss the matter at all. To many of them, spirituality seems "unscientific"—a topic for priests, rabbis, and philosophers rather than for scientists.

Cloninger was dissatisfied with standard methods of classifying personality because they were purely descriptive. They portrayed what personality looked like but not where it came from. He wanted a system that would reflect the underlying neurobiological and cultural sources of individual differences.

His solution was the biosocial model, which is woven from many separate threads of knowledge. Studies of twins, families, and adoptions were used to determine which personality traits are more genetic and which are more environmental. The stability of personality traits over an individual's life span was determined from longitudinal studies that tracked individuals over many years. Neuropharmacological and neurobehavioral studies suggested which brain chemicals were released and which structures activated when the various traits were exhibited.

Cloninger included self-transcendence in his biosocial model because he believed spirituality was an important part of life too long neglected by behavioral scientists. As he is fond of pointing out, each day more people pray or meditate than have sexual intercourse. Unlike Maslow, Cloninger is also a believer: a practicing Catholic. Ultimately, though, both of them had the same interest. They wanted to understand spirituality separate from organized religion and orthodoxy.

Cloninger consulted many different sources—Western and Eastern, historical and modern, religious and secular—to develop a composite picture of what it means to be spiritual. He scrutinized the lives of prophets and saints, mystics and seers, gurus and yogis. He read the humanistic psychologists' descriptions of self-actualizing people and the transpersonal psychologists' reports about meditators. Based on his research, Cloninger developed a scale that is based on three distinct but related components of spirituality: self-forgetfulness, transpersonal identification, and mysticism.

Self-Forgetfulness

Let me pose several questions to you. Do you ever get so involved with your work that you forget where you are or what time it is? Do you sometimes have the feeling that you are in the "zone" in terms of work, or sports, or music, and you can do no wrong? Have you ever been so in love with someone that you felt like there was no boundary between the two of you?

Most people have had this type of experience at least a few times in their lives. Spiritual people tend to have them more frequently. They score high on the TCI for self-forgetfulness, which is the first component of self-transcendence.

It's not easy to forget oneself. People think about themselves more than anyone else. Even when pondering a strictly impersonal question, such as a math problem, it's difficult not to think about how it might affect you personally. But sometimes a task or subject is so fascinating that you become utterly absorbed. Place and time become unimportant, personal concerns and problems fade away. Your concentration is focused, complete.

Such absorption can happen to a painter who becomes one with the process of painting, a musician who becomes lost in his music, a priest deep in prayer, or an ordinary person—such as an amateur gardener who is involved in planting, mulching, and weeding and suddenly discovers to his or her surprise that three hours have gone by. It is what psychologist Mihaly Csikszentmihalyi calls a "flow state." It can happen to anyone engaged in a challenging task that demands concentration and commitment, whether they're working, playing, enjoying a hobby, or taking part in a relationship. What's important, according to Csikszentmihalyi, is that the task demands commitment and provides immediate feedback, and that the individual's skill level is up to the challenge.

Self-forgetfulness means having this type of "flow" on a regular basis. People often experience flashes of insight or understanding when they are in this frame of mind. Creativity is maximized, originality is fostered. Even the most ordinary things seem fresh and new.

The downside of self-forgetfulness is that the person may be absent-minded or "out of it." The opposite is true for individuals who score low for this trait. They are too self-conscious to lose themselves in a task. They retain a strong sense of individuality even when working intensely, playing hard, or deeply involved in a relationship. They also tend to be more conventional, prosaic, and unimaginative than self-forgetful people.

Matteo Ricci, a sixteenth-century Italian missionary to China, epitomizes self-forgetfulness. Ricci, a Jesuit, played a key role in introducing the Chinese to both Christianity and Western science. Previous missionaries tried to convert the Chinese to Christianity by preaching in Latin and teaching them European customs. They didn't get far. Ricci thought it would be more effective to immerse himself in Chinese culture. He ate Chinese food, lived in a Chinese house, and dressed like a Chinese scholar. He learned to read, write, and speak the Chinese language.

Chinese is a difficult language. There is no alphabet, just 50,000 or so ideograms—a different character for every word. Ricci had to memorize each one. He didn't even have a dictionary; there weren't any. So Ricci developed a trick. He built a memory palace: a mental image of a series of buildings in which each word was represented by a specific image in a particular location. The word for war, for example, was symbolized by two warriors locked in combat in the southeast corner of a reception hall. In the northeast corner was the word for profit—a farmer holding a sickle, ready to cut his crops.

As Ricci's vocabulary expanded, he spent more and more time in his memory palace. He would forget about the real world for long periods of time as he wandered from word to word, image to image. There was no room for thoughts of himself, or indeed of any real person—only the Chinese characters and their mental pictures.

Ricci's self-forgetful study of Chinese paid off. Soon he could run through a list of several hundred Chinese characters and recall them both forward and backward. Over the course of twenty-seven years, he translated numerous biblical, scientific, and mathematical texts from Latin into Chinese, wrote a popular Chinese catechism titled "A True Doctrine of God," brought the first maps of Europe into the country, and in return sent home the first accurate maps of China. His converts were many.

Not all self-forgetful people achieve feats as prodigious as Ricci's. Many of them are spiritual, though, perhaps because of the way that self-forgetfulness facilitates the next two stages of self-transcendence.

Transpersonal Identification

I believe in God, only I spell it Nature.

—*Frank Lloyd Wright*

Are you concerned about protecting animals and plants from extinction? Do you feel a sense of unity with all the things around you? Would you risk your life to make the world a better place?

These are some of the questions used to assess the second subscale of self-transcendence, known as transpersonal identification. The hallmark of this trait is a feeling of connectedness to the universe and everything in it—animate and inanimate, human and nonhuman, anything and everything that can be seen, heard, smelled, or otherwise sensed. People who score high for transpersonal identification can become deeply, emotionally attached to other people, animals, trees, flowers, streams, or mountains. Sometimes they feel that everything is part of one living organism.

Transpersonal identification can lead people to make personal sacrifices to help others—for example, by fighting against war, poverty, or racism. It may inspire people to become environmentalists. Although there are no formal survey data, it is likely that members of the Sierra Club and Greenpeace score above average on this facet of self-transcendence. A drawback of transpersonal identification is that it can lead to fuzzy-headed idealism that actually hinders rather than helps the cause.

Individuals who score low on transpersonal identification feel less connected to the universe and therefore feel less responsible for what happens to the world and its inhabitants. They are more concerned about themselves than about others, more inclined to use nature than to appreciate it.

Love of nature is a recurring theme in spirituality, from the beginnings of civilization up to the present. "What else is nature but God?" asked Seneca the Younger, a Roman philosopher who lived 2,000 years ago. In the twentieth century, Albert Einstein put it this way: "Try and penetrate with our limited means the secrets of nature and you will find that, behind all the discernible concatenations, there remains something subtle, intangible, and inexplicable. Veneration for this force beyond anything that we can comprehend is my religion."

Love of nature is incorporated into many formal religions, especially in the East. Hinduism teaches love and respect for all living creatures, which is why vegetarianism is so common in India. Members of one sect, the Jains, go so far as to cover their mouths with cloth to avoid accidentally swallowing flying insects. The guiding principle of Taoism is to be in harmony with the original order of the universe; turning to nature is the way to discover that order. Mahayana Buddhism includes a vow to save all beings, while Shinto reveres the spirits manifested in foxes and squirrels.

Although Western religions are less consistently focused on nature, they are not insensitive to it. St. Francis of Assisi was a famous naturalist who preached to birds and tamed a wolf. His love for all aspects of the physical universe is seen by Christians as a kind of symbolic return to innocence experienced by Adam in the Garden of Eden.

One example of the principles of transpersonal identification is Albert Schweitzer—medical missionary, musician, theologian, and Nobel Peace Prize winner. Schweitzer, one of the most acclaimed self-transcendents of the twentieth century, developed a system of ethics that he called "reverence for life": that everything that maintains and enhances life is good, everything that destroys or hinders it is bad. The idea came to him while he was on one of many medical missions deep in the jungles of Africa.

Schweitzer believed that "all life is valuable and that we are united to all this life. From this knowledge comes our spiritual relationship with the universe." Even the simplest life-forms were sacred to him, and he admonished people to avoid injuring any living creature.

According to Schweitzer, all people possess an instinctive "reverence for life." But they are often reluctant to express it because they are afraid of being thought of as sentimental. To Schweitzer, expressing that reverence for life was an inner necessity that arises independent of thought or understanding. In other words, he believed that transpersonal identification is innate, not learned. It was a prescient thought.

Mysticism

> There is a central order to the universe, an order that can be directly apprehended by the soul in mystical union.
>
> —*Albert Einstein*

Have you often found yourself moved by a fine speech or piece of poetry? Do you sometimes feel a spiritual connection to other people that can't be explained in words? Do you think mystical experiences are just wishful thinking, or real?

These questions are from the third and final subscale of self-transcendence, which Cloninger calls "spiritual acceptance versus rational materialism" but that I refer to by the simpler term "mysticism."

Individuals who score high for mysticism are fascinated by things that can't be explained by science. They see a loaf of bread that resembles Jesus or a parking space that opens up just in the nick of time as evidence of a higher power. Often they claim to have a "sixth sense," or extrasensory perception. They believe in miracles.

Low scorers on this subscale are more materialistic and objective. They see an unusual loaf of bread or an unexpected parking opportunity as nothing more than coincidence. They don't believe in things that can't be explained scientifically.

Mysticism is the part of self-transcendence that most obviously relates to traditional spirituality and religion, because it includes belief in the supernatural. Indeed, many of the world's religions were founded by mystical individuals such as Siddhartha Gautama, Jesus, Muhammad, Yazid Taifur al-Bistami (Sufi Muslim), Mary Baker Eddy (Christian Science), and Joseph Smith (Mormonism).

But you don't have to be religious to be mystical. Scientists, too, can score high on this aspect of self-transcendence. Albert Einstein is a prime example.

Einstein was not conventionally religious. He rejected the orthodox Judaism of his parents at the age of twelve, disavowed the idea of a soul separate from the body, and was doubtful of an afterlife. Nor did he believe in the conventional image of God as a personal being who is concerned with our lives, answers our prayers, and judges us when we die.

Nevertheless, Einstein was a profoundly spiritual individual who believed that "the fairest thing we can experience is the mysterious." His reverence was directed to the harmony of the cosmos, the sheer wonder of existence. His God, who "didn't play dice," was "the grandeur of reason incarnate."

Cloninger believes that the psychological function of mysticism is intuitiveness. In support of this idea, he has found that people who score high on the mysticism subscale also score high for various measures of creativity. However, individuals who score high for mysticism but lack psychological maturity may be prone to psychosis.

Scientists often denigrate "intuitiveness" as little more than jumping to unsupported conclusions. After all, the ostensive purpose of science is to prove things, not to make guesses. Einstein felt otherwise. He never tried to separate his science from his spirituality. In fact, he believed the two were interconnected. As he commented in an interview that nicely encapsulated the connections between mysticism, creativity, and spirituality:

> The most beautiful and most profound religious emotion that we can experience is the sensation of the mystical. And this mysticality is the power of all true science. If there is any such concept as a God, it is a subtle spirit, not an image of a man that so many have fixed in their minds. In essence, my religion consists of a humble admiration for this illimitable superior spirit that reveals itself in the slight details that we are able to perceive with our frail and feeble minds.

Does It Hang Together?

Cloninger developed the self-transcendence scale based on the lives of spiritual people. Based on his criteria, Siddhartha Gautama, Mahatma Gandhi, Albert Schweitzer, Albert Einstein, and Tenkai would score high for self-transcendence. Genghis Khan, Queen Victoria, and Dwight Eisenhower probably would not. It seems like a logical way to measure spirituality.

But is it really? Is self-transcendence a valid psychological trait, or is it just a mishmash of various bits and pieces of personality?

To find out, Cloninger gave the TCI to ordinary people and then analyzed their responses to the questions about self-transcendence to see if their answers hung together in a coherent way. The statistical tool that he used to examine the data was factor analysis, one of the handiest implements in the psychologist's toolbox. It can be applied to individual items, like the questions in a personality inventory, or to groups of items, such as the subscales for mysticism, transpersonal identification, and self-forgetfulness. The aim is to determine whether different items are fundamentally related to one another. If so, they may be caused by a common source.

To give an example, suppose I collected information about people's eye color, hair color, height, and weight. Statistical analysis would reveal two factors: coloration and size. Hair and eye color tend to go together because they both use the same pigments; that's why most people with brown eyes also have brown or black hair. Height and weight go together because, given the same fat percentage, a taller person will simply have more mass than a shorter person. Of course, there would be some people who didn't fit the mold—short, heavy, brown-eyed blondes, for example—but they are the exception rather than the rule.

The same approach can be applied to personality. Cloninger's data came from three hundred people at a mall in St. Louis. There was nothing special about the individuals who participated. They were just your typical mix of shoppers: male and female, young and old, black and white.

First, Cloninger looked at how well the individual questions of the self-transcendence subscales hung together. The TCI contains 33 questions about self-transcendence: 11 for self-forgetfulness, 9 for transpersonal identification, and 13 for mysticism. To test for coherence, Cloninger analyzed how well people's answers to questions from the same subscale went together. For example, when testing for mysticism coherence, he asked whether those fascinated by the unexplained also sometimes think they have a "sixth sense." And when studying the coherence of self-forgetfulness, he asked whether people who often lose track of time are also frequently considered absentminded.

The answer to the question of whether or not these questions had coherence was a resounding yes. The reliability coefficients for each of the three subscales of self-transcendence were greater than 75. That means that all of the answers to the individual questions within each subscale had a far greater than random chance of hanging together. This part of the test was like asking people what their weight is and then actually weighing them on a scale; the answers were not exactly the same but were very close.

Next Cloninger turned to the three components that make up self-transcendence. Do they also cohere? In other words, do people who normally experience self-forgetfulness also score high for mysticism and transpersonal identification? Again the answer was a resounding yes. All of the intercorrelations were greater than 50, a highly significant degree of relatedness. (Correlations are measures of the relationship between variables. For the purposes of this book, the correlations are given on a scale of 0 to 100, where 0 means no relationship exists and 100 signifies an exact correspondence.)

Even more impressive, when factor analysis was applied to the entire TCI, the three subscales of self-transcendence were clearly separate from all other temperament and character traits. They formed their own distinctive coherence. In other words, self-transcendence is as distinct from other parts of personality as eye and hair coloration are from size.

These mathematical results show that there's something fundamentally similar about the three components of self-trancendence—self-forgetfulness, transpersonal identification, and mysticism.

Whatever it is that causes one of these traits to score higher or lower on the scale also makes the other components score higher or lower. There's a common root, a shared mechanism.

Of course, statistics alone can't tell us what that common root is. It could be the result of a gene. But it could equally well be the result of environment or culture. Math can tell us *whether* things hang together but not the reason *why*. For that we need other tools.

Cloninger's shopping-mall experiment had a practical implication: simplicity. One of the main motivations behind using factor analysis is to allow scientists to measure a trait with just a few questions instead of with many. To return to the example of measuring a person's weight, we might have first asked our subjects how much they weighed, then put them on a bathroom scale, then checked again with a laboratory balance. That would give us three separate numbers that all basically measure the same thing—three numbers to juggle when one would suffice. (Granted, some people will be mistaken about their weight, but then some scales are not that accurate, either.)

The same is true for spirituality. In subsequent experiments, in which we would be dealing with a multiplicity of genes and other factors, it was vital to measure self-transcendence as frugally as possible. Dealing with one number for the overall scale, or with three if using the subscales, would be reasonable. Handling thirty-three numbers, one for each question on the questionnaire, would have been impractical. Simple is good in science. Self-transcendence is, so far, the simplest way we have to measure spirituality.

Hershey Heaven

Alfred Day Hershey, who won a Nobel Prize for showing that DNA is genetic information, had a simple definition of heaven: "To find one good experiment and then keep on repeating it over and over again."

His comment might seem silly, but it emphasizes the importance of replication in science, especially in the behavioral sciences, where there always is a certain degree of subjectivity due to the very nature of the topic. Before starting my hunt for the genes that influence our tendency toward spirituality, it was important to see if Cloninger's shopping-mall results on self-transcendence could be repeated. Would self-transcendence hold up as a cohesive factor in other populations?

To find out, we analyzed the results of 1,001 subjects who took the TCI in connection with genetic studies conducted by the National Institutes of Health. Some of them were students I met at George Mason University. The rest came from other colleges in the Washington, D.C., area or were recruited through advertisements in local newspapers. They were young and old, male and female, black, Hispanic, and white. Some were religious, while others were not. They were, by and large, a pretty ordinary group of people, as we had hoped.

The first order of business was to check whether the individual questions used to construct the three self-transcendence subscales hung together in our data as they had in Cloninger's study. Indeed they did. The reliability coefficients were 66 for self-forgetfulness, 71 for transpersonal identification, and 80 for mysticism, all of which are quite respectable. If the various questions were unrelated, the scores would have all been zero. One or two questions didn't fit Cloninger's model, particularly the ones about self-forgetfulness, but they were in the minority.

Next we looked at the overall structure of self-transcendence. Did the three subscales hang together? Was there coherence? Again the answer was a resounding yes. The interscale correlations were all above 45, which is more than acceptable for this type of analysis. When we tried to force the data into other types of models, they didn't fit nearly as well; the results of Cloninger's study were confirmed.

The corollary to Hershey's definition of heaven is "There's no such thing as too much of a good thing." We performed an additional factor analysis of self-transcendence on 387 subjects for whom we had TCI data but not genotypes. Once again, self-transcendence stood out in terms of its coherence. It didn't matter which subjects we looked at or how we cranked the numbers. The answer came out the same every time.

A Woman's Place

Our reason for confirming the validity of the self-transcendence scale was that we wanted to measure individual differences in spirituality and correlate them with genes. It was also important, however, to know if there were any group differences in the scale, because these would complicate and possibly compromise the genetic analysis. Our database of 1,388 subjects allowed us to examine the relationship between self-transcendence and three potentially important demographic variables: race, age, and gender.

There were no significant differences in self-transcendence among different racial and ethnic categories. Although Americans of European, African, Hispanic, and Asian descent have varying religious traditions, their scores on the spirituality scale were all very similar.

There were also no consistent connections between one's age and self-transcendence. This was initially surprising, since Cloninger has speculated that self-transcendence is a maturational trait. (Nonetheless, his data show a similar lack of correlation.) Old and young alike have the capacity to be spiritual.

There was, however, a gender difference. Women scored 18 percent higher on self-transcendence than men. The difference was significant for both the overall scale and each subscale, and it held true regardless of age, race, and ethnicity.

Although gender is statistically related to many different personality traits, the effect on self-transcendence is particularly strong. The reason for this is unknown. A statistical analysis showed that the higher scores of women could not be accounted for by any of the other personality factors we measured. It might be something to do with the fact that women are more willing to express their feelings than men, or perhaps there's something about our society that brings out the spiritual in women. Or it might have something to do with their genes—a possibility we would later have a chance to test experimentally.

Not a Factor by Any Other Name

There are only so many fundamental personality characteristics that distinguish one person from the next. Many of them, like extroversion versus introversion, have been recognized for centuries. Nevertheless, each psychologist who develops a new system for personality classification has a seemingly irresistible tendency to rename each of the traits they "discover." It's only human to seek recognition, of course, but it can get confusing for those of us who study such things.

Cloninger is not immune to this tendency to rename things. The personality trait he calls "harm avoidance," for example, is basically the same as the trait Hans Eysenck termed "neuroticism," which in turn is virtually identical to what Raymond Catell called "anxiety." All three descriptions are based on related aspects of negative emotionality, such as worry, depression, and shyness. More important, all three traits correlate strongly when the Cloninger, Eysenck, and Catell questionnaires are administered to the same individuals.

Could self-transcendence, we wondered, be such a trait: an old trait under a new name? It didn't appear to be like anything else in the personality literature, but looks can be deceiving. A better test would be to administer several different personality inventories to the same subjects and then check whether self-transcendence is closely related to any previously described traits.

Fortunately, we had given almost all of our subjects two other personality tests as well: the NEO and the 16PF. The NEO breaks down personality into five basic traits called neuroticism, extroversion, openness, agreeableness, and conscientiousness. It's a widely used inventory that has been tested and validated in many different populations from around the world. The 16PF is an older test, devised by Raymond Catell, that envisions personality in terms of sixteen basic traits that make up five higher groupings: extroversion, anxiety, independence, tough-mindedness, and control.

We found that none of the factors or traits previously identified were the same as self-transcendence. Our statistical analysis clearly showed that both the overall scale and the three subscales of self-transcendence are unique. They are not contained in either of the other two personality inventories. The only possible exception was the openness factor, which contains a measure of imagination that is similar to mysticism. But even when all of the factors from the NEO were used to predict a person's self-transcendence score, the overall accuracy was less than 25 percent.

A Unique Trait

I apologize for all this number crunching, but it's a necessary stepping-stone in understanding the origins and the meaning of spirituality. What it shows is that spirituality, as measured by the yardstick of self-transcendence, really is a unique trait, not just a personality quirk.

This is important, because many nonbelievers see spirituality as nothing but an expression of underlying insecurity—a fear of death, for example. But if that were the case, we would have found a strong correlation between self-transcendence and measures of anxiety. We did not.

Others see spiritual practices such as meditation as nothing more than an attempt to stimulate a jaded mind—a sort of safe form of mind alteration, similar to drug use. If that were true, we would have found a strong association between self-transcendence and the personality traits of novelty and thrill seeking. But again, we did not.

Although statistics can't tell us what spirituality is or is not, or where it comes from, it can help to measure spirituality in individuals and confirm its uniqueness. In a study of the connection between genes and spirituality, that's a good place to start.

Three

An Inherited Predisposition

I myself believe that the evidence for God
lies primarily in inner personal experiences.
—*William James*

Unlike most young children, Jane and Rose loved going to mass, not just on Sunday but during the week as well. They both decided as teenagers to devote their lives to the church and took their vows together. Today, Sister Jane Frances and Sister Rose Marie are nuns at the same convent in Akron, Ohio. In addition to their mutual interest in God and spirituality, Jane and Rose have one other important similarity: their DNA. They are identical twins—the product of the same fertilized egg.

Twins, especially identical ones like Jane and Rose, are fascinating. There is something mysterious and alluring about people who look and sound identical.

But identical twins are more than curiosities. Because they share identical DNA, for scientists they offer a way to dissect the role of genes and environment in complex human characteristics like spirituality.

There are plenty of anecdotes about spiritually inclined twins such as Jane and Rose. One need only visit the annual twins convention held each August in Twinsville, Ohio, to hear them by the score. But are such stories the exception or the rule? How much of spirituality is inherited, and how much is environmental?

Galton's Legacy

The first scientist to recognize the potential of twins to answer quantitative questions about the origins of behavior was Sir Francis Galton, a nineteenth-century English scientist who was a cousin of Darwin. A child prodigy, adult polymath, explorer, geographer, statistician, and psychologist, Galton is best remembered for originating the twin experiment that has been the mainstay of behavior genetics to this day.

The main use of twins in behavior genetics research is to determine heritability, which is defined as the percentage of variation in a behavior that is due to genetic differences. Heritability can be most directly measured by comparing identical twins who were separated at birth and raised apart. Because such twins have the same genes but are raised in different environments, the extent to which they are similar to each other is a direct approximation of heritability. The degree of similarity can be calculated as a correlation.

Unfortunately, there is a serious problem in studying identical twins separated at birth. There just aren't enough of them. Fewer than two hundred pairs are known in the entire United States, and the number is getting smaller all the time, as it is now customary for twins to be adopted together. The alternative for research purposes is to compare identical and fraternal twin pairs who were reared together. Since fraternal twins, like ordinary siblings, share only half of their genes, they should be less similar to each other than identical twins to the extent that genes are important.

Although twins reared together share the same environment, their environment can be subtracted from the equation, since it's the same regardless of zygosity—that is, whether the twins develop from the same or different eggs. By comparing the correlations for the two types of twins, fraternal and identical, it is possible to calculate how much of the individual differences in a trait is due to genes and how much is due to growing up in the same or different households.

Galton invented the twin method of research long before anybody knew what genes were or how they worked. In fact, he didn't even know why some twins appeared so similar and others did not. But he guessed—correctly, as we now know—that it was because the identical twins had the same inherited makeup, whereas the nonidentical twins were like ordinary brothers and sisters. Galton therefore reasoned that "their history affords a means of distinguishing between the effects of tendencies received at birth, or those that were imposed by their circumstances of their after lives; in other words, between the effects of nature and nurture."

Galton used his new twin method to examine many different aspects of human development, health, and behavior, including toothaches, various diseases, personality, and life events. He even explored whether or not dogs can smell the difference between twins.

Galton also was fascinated by spirituality and religion. He was the first modern scientist to try to objectively study the efficacy of prayer and belief. To do so, he compared well-known theologians and clergy to other eminent men. He found that the life spans of the clergy were no longer and their illnesses no less frequent or severe than those of the laypeople.

The one thing that Galton did not do was to combine his interest in twins with his curiosity about spirituality to explore whether or not belief has an inherited component. To ask the question properly, he would have needed both an appropriate measuring stick and a sizable collection of twins who had been reared apart. Those two elements didn't coalesce until nearly a century later in the American Midwest.

The Minnesota Experiment

Realizing the potential importance of twins reared apart for behavior research, twenty years ago a group of scientists at the University of Minnesota began to systematically track down pairs who had been separated at birth for one reason or another. Some of these twins had been reunited in early adulthood; others did not even know they had a twin until middle age or later. Some twins met by complete accident, such as the two men who ran into each other at a gay bar; others had spent years looking for each other. Although there were pairs who were reared in quite similar environments, in some cases by relatives, there were others who were brought up under strikingly different circumstances. Oscar Stohr and Jack Yufe, for example, were raised in different countries and disparate religions—Oscar as a Catholic in a small German-Czech border town, where he joined a Hitler Youth group, Jack as a Jew in Trinidad, Venezuela, and then Israel, where he lived in a kibbutz on the Sea of Galilee.

Although separated identical twin pairs are few and far between, the Minnesota scientists eventually accumulated enough pairs to begin to ask quantitative questions about complex human characteristics. In the first published study of its kind, the researchers examined 53 pairs of identical twins and 31 pairs of fraternal twins, all reared apart, for five different scales of religiousness. The emphasis of the study was on the sorts of orthodox values, beliefs, and practices that are promoted by typical organized religions. Some of the scales looked at how important religious faith was to the twins' lives. Others asked how much of their leisure time they spent in religious activities, such as attending services or working for their church or synagogue. There were questions about interest in religious occupations, such as working as a priest, minister, rabbi, or missionary; others tapped into particular tenets, such as belief in God.

The results of their study were consistent. For every scale examined, genes seemed to play an important part. The calculated heritabilities were all between 41 and 52 percent, meaning that genes were responsible for roughly half of the variation in religiousness from one twin to the next. In other words, the study seemed to suggest that at least part of the reason people believe that religion can help to answer life's questions is their DNA.

Although this first study provided fascinating new information about religiousness, it didn't really address one's inherent spirituality. No measure of spirituality, such as the self-transcendence scale, was included in the analysis. In fact, little attention was given to people's underlying motivations and feelings. In a second study, however, the Minnesota researchers did ask about religious motivation—about the *reasons* that people become religious—which is at least closer to spirituality.

The Minnesota researchers analyzed religious motivation by using a questionnaire that distinguishes intrinsic from extrinsic religiousness. Intrinsically religious people live their religion. They often feel the presence of God, they pray as often when they are alone as at their place of worship, and they try to practice their beliefs in every aspect of their lives. Extrinsically religious people go to church or synagogue to see their friends or make new ones; they are even willing at times to suppress their religious beliefs to impress others. Intrinsic and extrinsic religiousness are unrelated; the correlation between the two scales is zero.

When the twin data was analyzed, the results for extrinsic religiousness were ambiguous. There was a significant correlation for identical twins raised apart, suggesting that genes might play a role. But the correlation for fraternal twins was actually greater, which doesn't make any sense for a heritable trait. The numbers were just too inconsistent to interpret.

The results for intrinsic religiousness were clearer. This measure, which comes closest to spirituality in the Minnesota study, was found to be substantially genetic. The correlation for identical twins was 37, about double the correlation of 20 for fraternal twins. When these numbers were analyzed, the estimate for heritability came out to 43 percent. In other words, nearly half of the reason that the twins felt religion helped them, spent time privately praying, and had a sense of God's presence was inherited. Since these twins were raised by different parents, in different neighborhoods, and sometimes even in different religions, their similarities seemed to be the result of their DNA rather than their environment. Something in their genes helped to push them toward religion.

Self-Transcendence and Heritability

It's difficult to imagine two scientists as dissimilar as Nicholas Martin and Lindon Eaves. Martin is tall and lean, extroverted and loud, an Australian and proud of it. Eaves is short and plump, a shy and soft-spoken Briton transplanted to the United States. Martin, who trained as a psychologist, is an atheist. Eaves, a geneticist and mathematician, wears the collar of an Anglican priest and preaches every Sunday. But there are two things that both men love: twins and pushing the envelope.

Martin established the Australian Twin Registry, a large and systematic collection of identical and fraternal twins that has been used to study everything from sexual orientation to freckling. Eaves's twin population, the Virginia 20,000, is even larger and has been the subject of hundreds of research papers.

In 1999, Martin and Eaves teamed up with Australian scientist Katharine Kirk to pursue a new topic for behavioral genetics: self-transcendence.

The sample was drawn from the Australian National Health and Medical Research Council Twin Registry, a volunteer register begun in 1978 that has a total of about 25,000 pairs of twins of either zygosity and various ages. The self-transcendence study focused on twins over 50 years of age, who were mailed a health and lifestyle questionnaire. There were 3,116 replies, 1,279 from complete pairs and 558 from singles, which represented a response rate of 71 percent—a typical result for a mail survey. The subjects had an average age of 61, with a range of 50 to 94 years, and varied educational and socioeconomic backgrounds. There were about twice as many women as men, which again is typical for this type of research.

Self-transcendence was assessed by fifteen questions selected from Cloninger's TCI inventory. Although each of the facets of self-transcendence was covered by the abbreviated test, there were not enough items to determine individual scores for self-forgetfulness, mysticism, and transpersonal identification.

Just as Cloninger found in St. Louis and we confirmed in Bethesda, the various questions about self-forgetfulness, transpersonal identification, and mysticism all hung together in a coherent fashion. Furthermore, the scale reliability and factor analytic statistics were virtually identical to those found in the United States. (The one exception was the item "I love the blooming of flowers in the spring as much as seeing an old friend again," which for some unknown reason didn't fit in with the rest of the questions in Australia. Go figure.)

The twins were also asked several questions about their religious affiliation and church attendance. This allowed the researchers to distinguish between spirituality, as measured by the self-transcendence scale, and more traditional religiousness. The study included questions about health, anxiety, depression, optimism, and various aspects of personality.

How did the researchers conduct their studies? First, they evaluated the data, using a modeling technique that took into account three main sources of variation in self-transcendence: genetic influences, shared environmental influences, and unique environmental influences. The first two factors make twins alike; the third makes them different. The analysis indicated that genes are responsible for 48 percent of the variation in self-transcendence in twins, both male and female. The remaining 52 percent of variance was due to environmental factors for females. Age also had an effect; in males, it accounted for 4 percent of variance. (Environmental factors accounted for the other 48 percent.)

The researchers also examined the data by a statistical technique called "multivariate modeling." Once again, they found that genes played an important role in self-transcendence. Using this analysis, the estimated heritabilities were 37 percent for men and 41 percent for women, which are similar to the numbers obtained in the first analysis. The take-home lesson from the Martin and Eaves study was clear: Genes are a major factor in self-transcendence. In other words, spirituality is, in good measure, an inherited trait.

To test the role of genes in self-transcendence, the scientists compared the similarity of responses in identical twins to fraternal twins.

There was a striking difference.

The self-transcendence scores of identical twins were more alike by far than those of fraternal twins. For males, the correlations came in at 40 in monozygotic twins, compared to 18 in dizygotic twins; for females, the corresponding numbers were 49 and 26. In other words, the ratio was close to 2 to 1 in both sexes, which is just what one would expect for a genetically mediated trait, since identical twins are twice as similar at the DNA level as fraternal twins.

The Nature of Nurture

The main purpose of twin experiments is to determine to what extent genes are important for different traits—to get at the nature side of the "nature versus nurture" question. But twin studies are equally informative with regard to the role of the environment as a factor—the nurture side of the story. Not only do they let us know that nurture is important, they can tell us what type of nurture makes a difference.

"The environment" is actually a bit of a misnomer, since the factors that affect who we become are certainly not a single entity. The environment includes everything from the type of diaper you wore as an infant and the weather on your third birthday to your parents' income and the amount of lead in your classroom paint. It's the catechism you were taught in Sunday school and the TV show you watched last night. The environment is anything and everything—biological, physical, intellectual—that you didn't inherit as DNA.

Behavioral scientists break down the environment into two major categories: shared and unique. In twin studies, the shared environment consists of everything that both twins experienced by growing up in the same household. This includes general parenting style, income level, social class, schooling, and religious upbringing.

The unique environment is everything else. It includes each and every factor and experience that twins do not share. If the twins have different schoolteachers, that's a unique component of the environment. If one of them gets measles and the other does not, that, too, is unique.

Is shared or unique environment more important in terms of spirituality? Twin data can be used to answer this question. If the shared environment were important, for example, both identical and fraternal twins would be more similar to each other than random individuals who grew up in different households. Furthermore, the extent of correlation due to shared environment would not be influenced by their genes, since both types of twins would have shared it to the same extent. As a result, fraternal twins would resemble each other more than expected on the basis of their 50 percent genetic similarity.

By contrast, suppose that the unique environment were the critical factor. If that were true, fraternal twins and identical twins would be less similar to each other, rather than more similar. Whatever similarity existed between twins would be due to genes plus shared environment, and whatever part of that similarity is inherited would be twice as great in identical twins as in fraternal twins. By combining these two facts, it is easy to derive equations that parse out the separate contributions of genes, shared environment, and unique environment.

An important assumption in this calculation is that identical and fraternal twins share environment to the same extent. Critics of twin experiments often question this assertion, pointing to anecdotes of identical twins being treated very similarly. Fortunately, it's possible to test the "equal environments assumption," as it's known, by an experiment. The trick is to take advantage of the fact that some parents mistake identical twins for fraternal and vice versa. When such "pseudo-identical" and "pseudo-fraternal" twins are compared with real identical and fraternal twins, there's no difference. It doesn't make any difference whether parents think that the twins are genetically the same or not. It only matters whether they have the same DNA.

When this type of mathematical analysis was applied to the twin data on self-transcendence, the result was clear. What is important is unique environment. In the univariate statistical model, the unique environment accounted for 48 to 52 percent of variance in men and women, respectively, and shared environment accounted for zero. In the multivariate model, unique environment accounted for 42 to 50 percent of variance, and shared environment remained insignificant.

This was a surprising result. The implication is that spirituality, at least as measured by self-transcendence, doesn't result from outside influences. Contrary to what many people might believe, children don't learn to be spiritual from their parents, teachers, priests, imams, ministers, or rabbis, nor from their culture or society. All of these influences are equally shared by identical and fraternal twins who are raised together, and yet the two types of twins are strikingly dissimilar in the extent to which they correlate for self-transcendence. In other words, William James was right: Spirituality comes from within. The kernel must be there from the start. It must be part of their genes.

The Difference Between Spirituality and Religiousness

When Galton first studied twins more than a century ago, the results were so striking that he feared nobody would believe him. "My only fear is that my evidence seems to prove too much and may be discredited on that account, as it seems contrary to all experience that nurture should go for so little," he wrote.

The twin researchers in Minnesota and Australia had the same concern. Their results seemed almost too good to be true. It seemed that everything they looked at was heavily genetic and that the shared environment didn't make a bit of difference. This appeared to be true not just for religious motivations and self-transcendence, as in the studies described above, but also for such basic traits as intelligence, personality, activity level, mental disorders—including schizophrenia and manic depression—and assorted behaviors ranging from cigarette smoking to petty criminality. And then there were the truly eerie similarities between the twins raised apart, like those who gave their dogs the same name or used the same obscure brand of toothpaste.

Critics of behavior genetics were incredulous. Could there really be "a gene for" something as idiosyncratic as one's preference in dog names or brand of toothpaste? Perhaps, said the skeptics, there was something fundamentally wrong with the whole twin approach. Maybe there was some basic error in the methodology that made everything seem genetic even when it was not. And why did the shared environment seem so unimportant? Perhaps twin experiments were methodologically incapable of sensing the role of parents, teachers, and society.

The best way to refute the critics would be to use the twin method to look at a trait that was at least partially learned—something that was influenced by culture as much as heredity. With that idea in mind, Martin and Eaves decided to look at religious-service attendance in the same sample of Australian twins whom they had studied for self-transcendence. Their reasoning was that the frequency with which a person goes to church, synagogue, or mosque is more likely to be learned than inherited, because it varies so much from one culture to the next. For example, Australians go to church much less frequently than do Americans. About half of the folks down under, compared to less than one-third of people in the States, attend rarely or never. And only 7 percent of Australians go to church more than once a week, whereas the number in America is more than twice that. Since Australians and Americans have the same genes, church attendance is more likely to be learned than inherited.

That is exactly what analysis of the Australian twin data showed. The main factor that caused twins to be similar to each other for church attendance was shared environment, not genes. The correlations were almost identical for identical and fraternal twins, allowing the researchers to conclude that shared environment was responsible for 43 percent of the variance in both males and females; the remaining variation was due to a mixture of unique environment and a limited genetic component.

Could the same genes be responsible for the inherited components of religious attendance and spirituality in women? To test that possibility, the researchers looked at the correlations between church attendance in one twin and self-transcendence in the co-twin. If the same genes were involved in both traits, these cross-twin, cross-trait correlations would be stronger in identical than in fraternal twins.

This was not the case for church attendance and self-transcendence. They aren't linked at all. Both the within-twin and across-twin correlations were small and were completely independent of zygosity. Whatever the genes are for spirituality, they don't have any effect on how often people go to church.

Maslow, Cloninger, and many others before them and since have argued that spirituality and religiousness are fundamentally different. The twin studies, by looking quantitatively at both qualities in a single population, strongly support the distinction. More important, they tell us something about why they differ. Religiousness, as measured by church attendance, is learned in the classical sense—from parents, teachers, religious leaders, and peers. People go to church or mosque or temple because that's what they were told to do. Spirituality, as measured by self-transcendence, is more innate. It comes from within, not from without. Of course, spirituality has to be developed, just like any other talent. But the evidence suggests the predisposition is there from the beginning.

Brothers and Sisters

Although studies of identical twins like Jane and Rose play an important role in figuring out whether heredity is important for a trait, they're useless for actually isolating the genes. The DNA of identical twins is too similar to be informative in mapping experiments. To properly conduct such studies, what's needed are fraternal twins or siblings—individuals different enough for genetic localization (since 50 percent of their DNA variations are distinct), but similar enough to compare (since the other half of their DNA variations are identical). Siblings also usually are raised in the same environment.

Since siblings are much more common than fraternal twins, we focused our attention on them. As I described in the previous chapter, we recruited pairs of brothers and sisters from local colleges and the community, took some cells to isolate DNA, and gave them the TCI to measure self-transcendence. Before going any further, though, we needed to confirm the twin results about heritability in our population.

The logic was that siblings, who are 50 percent genetically related, should have similar scores for self-transcendence if it is indeed a heritable trait. If sibling scores were not related—if the correlation were zero—then self-transcendence could not be heritable in our population. It is important to emphasize that this test cannot *prove* heritability, since a positive correlation could, in principle, be due to genes, shared environment, or a combination of both. But it can be used to double-check that the population is consistent with previous studies of heritability, since a negative result would mean genes could not be involved.

When we examined our group of 447 pairs of siblings, they indeed scored similarly for self-transcendence. The correlation was 37, which was as high or higher than expected from the heritability calculated for Australian twins by Martin and Eaves.

The sibling correlation for self-transcendence was true regardless of sex, age, or race. Brothers were like brothers, sisters like sisters. Older siblings were as alike as those still in college. Regardless of race or culture, if one member of a pair scored particularly high or low for self-transcendence, his or her sibling usually did, too.

These results don't mean that siblings are *always* the same. There are plenty of exceptions, which is to be expected, since although siblings share 50 percent of their DNA variations, the other 50 percent are different. And even though siblings grow up in the same shared environment, their unique environment is just that—unique.

Sometimes the differences between siblings are striking. Tenkai, the Zen monk I describe in Chapter One, has a brother who is an executive in a manufacturing firm in Germany. He thinks spirituality is nonsense and laughs when Tenkai explains why he meditates. He has little interest in social justice or nature; he hates the Green Party with a passion. These two siblings couldn't be farther apart on the self-transcendence scale.

But there are also siblings like Gloria and Louise. I met them one fine spring Sunday morning at Mount Gilead Baptist, a predominantly African American church located a few blocks from my house in a gentrified neighborhood of Washington, D.C., that is now predominantly white. In attendance were ostrich-plumed women of a certain age with their grandchildren in tow, white-garbed church ladies with their dark-suited husbands, a sprinkling of young couples.

Gloria has been going to this church her entire life. She attended Sunday school when her parents still lived in the neighborhood, she was married here, and now she's a deaconess and a good friend to most all of the parishioners. But her devotion to the church goes beyond the social. She prays regularly—not just in the morning or evening, but repeatedly through the day. She has found Christ—not once or twice but over and over. She is a believer.

This morning it is Gloria's duty to read the announcements from the pulpit. She exhorts her flock to publicize an upcoming church picnic, saying, "The best way to get the news out is to telephone, telegraph, or tell-a-woman." With the last comment she looks directly at a woman in the audience and gives a hearty chuckle.

The woman is her sister, Louise, the black sheep of the family. Louise never liked Sunday school. She didn't enjoy regular school, either, and dropped out when she became pregnant with the first of four children at age sixteen. For the next twenty-five years, Louise struggled with alcohol, drugs, and a string of "good-for-nothing" men who seemed to be mostly interested in her welfare checks.

Then, as the result of a twelve-step program, Louise found God. She discovered that faith in a higher power was the one thing that could keep her off drugs and alcohol, and she developed a spiritual program that involved regular prayer and helping others. Louise, too, became a believer.

What makes some siblings, like Tenkai and his brother, so spiritually dissimilar despite their common upbringing? And what makes others, like Gloria and Louise, so similar despite their very different life trajectories? Could it be something in their genes? There was only one way to find out.

Four

The God Gene

There is very little difference between one person and another,
but, what little there is, is very important.
—*William James*

Have you ever seen your DNA? You can, you know. Here's a simple recipe that you can use to extract your genetic material. Start out with some cells; a few drops of blood or a tablespoon of spit will do. Crack them open by adding detergent. We use pure sodium dodecyl sulfate in the laboratory, but most drugstore shampoos work nearly as well. Next, remove the proteins by adding table salt until a large cloudy precipitate appears; pour the precipitate through a coffee filter. To the clear filtrate add four parts of vodka and place in the freezer.

Within an hour or so, the DNA will appear as a web of silky white threads. These can be twirled onto a glass rod, such as a martini stirrer, dried with a hair dryer, and dissolved in a glass of water.

That's your DNA. It's not much to look at—just a clear liquid when it's in solution. Yet within it lies the code that makes you a human instead of a chicken or a bacterium or a roundworm. The blueprint for your liver and your pancreas and your eyes and your hair. The instructions for the development of your brain from a few meager cells to the most complex biological structure in existence. And, at least in part, the instructions for your sense of spirituality.

How can something so seemingly simple be responsible for so much?

The Information Molecule

The secret of DNA lies not in its physical properties, which are quite ordinary, but in the information it carries. DNA molecules are like computer chips. They all look pretty much the same, yet they have vastly different functions depending on how they are programmed.

The information in DNA is stored in the form of building blocks known as bases. There are just four of them, abbreviated as A, G, C, and T. The bases punctuate the long string of the DNA molecule at regular intervals, like beads on a necklace. Each DNA molecule consists of two strings of beads, wound around each other in the double helix that James Watson and Francis Crick made famous in 1952. The order of the beads is not random. Every time there is an A on one strand, a T appears on the other, and every time there is a G on one strand, it is matched with a corresponding C. These base-pairing rules are what allow the DNA to be faithfully copied when cells divide.

The information content of DNA derives from the order of the bases. Every three bases specify one amino acid, a building block of a different type of molecule called a protein. For example, the DNA sequence ATG specifies the amino acid methionine, whereas the sequence GTA is read as valine. This is why the exact order of the bases is so important; methionine and valine are completely different amino acids, even though they are encoded by the same three bases in inverted order. There are twenty different amino acids, and they are chemically much more diverse than the four bases. The conversion of DNA information to protein information involves the same two steps, transciption and translation, in every organism.

Amino acids are so important because they determine the structure of proteins, the key players in every biological activity. Proteins are scaffolds, the structures that make each cell and organ distinct. Enzymes, the catalysts that direct every reaction in living cells, are proteins. Hormones, the molecules that make us male or female, tall or short, sleepy or wide awake, are made of proteins. Neurotransmitters, the signals that tell brain cells what's going on, are proteins. We are proteins—and what proteins we are made of depends on exactly what DNA we have.

From the human genome sequence we know that our DNA has roughly 35,000 different genes, each of which codes for its own distinct protein. That's a surprisingly small number, far fewer than previous estimates of 50,000 to 150,000. Humans, chimpanzees, dogs, and mice have nearly the same number of genes, the common fruit fly has 14,000 genes, the flatworm has 18,000, and the plant *Arabidopsis thaliana*, a type of weed, has a full 25,000.

It may seem surprising that we have only 10,000 more genes than a weed, but it's enough. You cannot tell the complexity of an organism from the number of genes it has any more than you can judge the sophistication of a software program from the number of lines of code it consumes.

Every human being has pretty much the same DNA, the same 35,000 or so genes. But "pretty much the same" does not mean identical. There are subtle differences from one person to the next. These variations, which are sometimes called polymorphisms, occur approximately once every 1,000 bases between unrelated humans. Since each human contains a total of about 3 billion bases of DNA, that means that there are about 3 million differences between your DNA and mine. These are what account for all of the inherited differences between us.

While one out of a thousand may not seem like much variation, differences between human and chimp DNA occur only once every 100 bases (or about ten times as often). Even with mice we differ at only one out of 30 bases. Our genomes are so similar because many proteins have the same biochemical function in every life-form.

Of the 35,000 genes present in the human genome, we know the function of only about one-third. These genes code for well-known proteins such as the globins that carry oxygen in the blood, the crystallins that make up the lens of the eye, and the gut enzymes that digest our food. Another one-third of genes have nonhuman homologues, meaning there are similar genes in other species. We don't know what these do yet, but we expect to soon since they can be manipulated in experimental animal systems. The remaining one-third of genes—more than 10,000 of them—are complete unknowns. We know they make proteins, but we have no idea what those proteins are or what they do. They are a mystery.

Because the function of so many human genes remains unknown, we still cannot deduce much by looking at an individual's DNA. We can tell an individual's sex and his or her traditional racial classification with reasonable certainty, but that's about it. We cannot, for example, determine a person's height. Although we know some of the genes involved in height, like the one that codes for growth hormone, most of them still have not been identified.

It's no surprise, then, that we can't simply look at the genome sequence and say where the God genes—the genes that create a predisposition to spirituality—are. Even if we knew the biochemical function of all the genes, we would not know how they interact with one another, and with the environment, to mold a trait as complex as spirituality.

What we can do, however, is identify sequences of DNA involved in the differences in spirituality observed from one person to the next. That is, we can look for what James so aptly called "causes of human diversity"—not the reason that all humans have some aptitude for spirituality, but the reason that some have more or less than others. We can attack that question without knowing the function of all the human genes by comparing people with different levels of spirituality. By analyzing their DNA, we can identify any sequence variations that track along with the strength of their beliefs, which for the purposes of our study are measured by the self-transcendence scale. All that we need to determine this are DNA samples from a series of subjects with known self-transcendence scores and a list of reasonable genes to look at.

Ordinary People

Tim and Peter Mooreland, who were recruited through our program at George Mason University, were typical subjects in our study. (Their names and several identifying characteristics have been altered.)

Tim is a twenty-one-year-old senior majoring in economics and business administration. He volunteered partly for the $40 we paid participants, partly because he became interested in genetics after following the biotech market for a business class. Tim has a solid B average, plays intramural basketball, and likes to have a few beers on the weekend. He smokes cigarettes, nearly a pack a day, but doesn't take any drugs.

Although Tim was raised Catholic, he no longer attends

church, except on Christmas and Easter with his family. Nevertheless, his self-transcendence score on the TCI was high—in the upper third of all males. He answered affirmatively on all of the statements about love of nature and feeling part of the universe. He admitted he would make personal sacrifices to make the world a better place. He is interested in things that can't be explained scientifically.

Peter, Tim's older brother, is twenty-seven and works for a Washington consulting firm. His days are spent at the firm, where he hopes to advance to a junior management position. He spends his nights and weekends with his fiancée. He also likes to play rugby with old college chums on the weekends. He drinks moderately, doesn't smoke cigarettes, but does indulge in a joint now and then.

Peter thinks religion is a good thing. Although he no longer attends church regularly, he intends to resume when he has children. His score on the self-transcendence scale, however, was considerably lower than his brother's—in the lower, not the upper, third. He likes nature but doesn't love it. He's interested in extrasensory perception but doesn't believe in miracles or the supernatural. On the survey statement "I often feel a strong sense of unity with all the things around me," he marked false.

Peter and Tim are pretty regular guys. Although they differ in their spirituality, their differences are not all that obvious from chatting with them. Neither of them is a Schweitzer or an Einstein—nor is either a Hitler or a Jack the Ripper, which suited our purposes perfectly. Our purpose was to understand more about everyday spirituality—not the spiritual extremes.

Tim and Peter actually were recruited for a study on cigarette smoking, rather than spirituality. They were part of a study of the genetics of smoking behavior and related personality traits sponsored by the National Cancer Institute. We gave them the TCI because it makes specific predictions about which brain chemicals and genes are involved in addictive personality characteristics, rather than because it has a self-transcendence scale. The spirituality aspect of the study was an unanticipated bonus, not anything planned or supported by my employer.

Other subjects were recruited as part of a similar study conducted in collaboration with the National Institute of Mental Health. The focus of that project was personality and mental health. Once again there was no specific focus on spirituality, nor any selection on the basis of self-transcendence levels; the NIMH is interested in mental disorders, not spirtuality, and the self-transcendence scale was included in the study purely because it is part of the TCI.

As a result, we had no prerequisites about spirituality for those admitted to the study. The one requirement we *were* strict about was that each pair consist of two same-sex siblings willing to participate. The reason for focusing on siblings is that they are perfect controls for each other. Because they are raised together, they have the same shared environment, including socioeconomic level, school system, religious training, and parents. And because they share half their DNA variations, they are also genetically matched, which makes it easier to follow the effect of any one particular gene. Most crucial, siblings, by definition, have the same racial-ethnic background since they have the same set of parents. That's important because different racial-ethnic groups often have different frequencies of certain DNA variations, just by chance, and this can lead to misleading associations. Comparing siblings to each other rather than to unrelated individuals avoids this potential ambiguity.

In all, we used 1,001 subjects, 623 from the cigarette smoking study and 378 from the personality study. Because there can be significant gender differences for self-transcendence scores, it was important to us that we had both men and women. We had 328 males and 673 females in the study, enough to look at the sexes separately. The greater number of females compared to males is typical for this type of study, because women generally are more likely than men to volunteer—perhaps because they are more self-transcendent.

The racial-ethnic composition of the population was determined by asking the volunteers to classify themselves on a standard checklist. The breakdown of the people in our study was 74.3 percent white, 6.9 percent Asian/Pacific Islander, 4.5 percent Hispanic/Latino, 9.3 percent African American/black, 0.2 percent Native American/Alaskan, and 4.5 percent mixed or other.

The average age of the participants was 32, with a range from 18 to 83. Two-thirds of them had college degrees; the remainder was split evenly between those with a high school education and those with an advanced degree. Personal income levels varied anywhere from less than $1,000 a year to more than $100,000 a year.

All in all, it was a pretty typical slice of America, although it was not a strictly population-based sample, which requires special recruitment strategies that are difficult for a volunteer study to achieve. But it also was not a deliberately peculiar population, which is common for many psychological and psychiatric studies that depend on patient populations. The individuals included in the study were just ordinary people.

Candidate Genes

The next question we had to tackle was what genes to look at.

It might seem that the ideal solution would be to determine everybody's complete genome sequence—all 3.2 billion bases of

DNA. But there are two problems with that approach. First, it can't be done with current technology. Just determining one genome sequence—which actually is a composite of several people's DNA—required more than ten years of research by dozens of large laboratories around the world. Repeating that Herculean task on the more than 1,000 individual DNA samples in our database would have been impossible. Someday we will have the technology to read out a person's DNA like a bar code on a can of peaches at the supermarket. But we can't do it yet.

The second problem is statistical rather than practical. The greater the number of different genes that are examined, the less likely that any particular association will be genuine. At first, this seems paradoxical. Shouldn't gathering more information make the results more reliable, not less? But consider the analogy between searching for a gene and flipping a coin.

Suppose you toss a coin 100 times and get 60 heads. You should be suspicious about that coin. There's only about a 5 percent probability of getting that many heads just by chance, which means there's a 95 percent probability that the coin is unbalanced. This is the usual, arbitrary cutoff that's used to determine whether a result is significant or not. But suppose you toss 20 different coins 100 times each. Now, getting 60 or more heads in one of them is not all that surprising. There's a 36 percent probability of such an event occurring just by chance, which means only a 65 percent likelihood that the coin is rigged. If you tossed 1,000 different coins 100 times each, at least one of them would almost certainly give 60 or more heads. It wouldn't mean that the coin was rigged—just that you are spending too much time flipping coins.

It's the same with genes. If you test only one or a few genes and find that one of them is associated with a trait, it's probably real. But if you test a large number of genes, there is a greater chance that any particular association is just a statistical accident. As described above, there are approximately 3 million polymorphisms in the human genome. That's a lot of coins to toss. If we compared complete genome sequences in our study, most of the apparent associations probably would be just by chance. So, before even starting the search for God genes, it was important to decide on a list of reasonable candidates. But where to begin?

The usual route, which entails starting from what's known about the brain chemistry or anatomy of the trait, didn't look hopeful. To do chemistry, you need pure molecules and an assay (analysis). We had neither. To do anatomy, you need to know the precise part of the body where the trait is expressed. Again, we were clueless, since the exact part of the brain involved in spirituality remains elusive.

Fortunately, we did have two hints about where to hunt for genes that influenced spirituality. The first was from pharmacology. Although there are no drugs known to directly influence spirituality, there are several that seem, at least superficially, to enhance or mimic the altered states of consciousness at the core of mysticism and self-transcendence. All of these drugs act on a related set of brain chemicals, the monoamines (such as serotonin and dopamine). I'll describe these drugs and the functions of monoamines in emotion, cognition, and consciousness in detail in the next chapter. The important point is that they provided a hint about what genes to consider as candidates in effecting self-transcendence.

The second hint came from the work of David Comings, a psychiatric geneticist at the City of Hope National Medical Center. Comings is best known for his controversial proposal that the same set of genes underlies a wide variety of behavioral conditions, including alcoholism, depression, conduct disorder, learning disabilities, post-traumatic stress disorder, stuttering, attention deficit disorder, and Tourette's syndrome. According to Comings, the same genes have beneficial effects, as well, such as promoting scientific curiosity.

To further understand the role these hypothetical genes may play in normal behavior, Comings performed a candidate gene survey for the seven traits measured by the TCI, including self-transcendence. He found significant associations between self-transcendence and seventeen different genetic variations. Interestingly, six of these associations—the largest number in any single category—occurred in genes dealing with monoamines, the very genes that had already attracted our interest.

Although Comings's results provided a tantalizing hint about where to look for spirituality genes, there were two problems with the design and interpretation of his study. One was the subject population. Only 204 subjects were involved, which is quite small when looking at such a complex trait. The study also consisted entirely of white males, which means it could tell us nothing about women or minority groups. Of even greater concern, more than half of the subjects were addicts from a Veterans Administration hospital addiction-treatment unit. Such a skewed study population badly confuses the analysis, since drugs and alcohol can radically alter a person's personality.

The second problem with the Comings study was that it fell into the "too many coins tossed" trap. He examined 59 different genes and 7 different traits for a total of 413 comparisons. There were just too many genes, too many traits, and too few subjects for any one of the associations to be completely believable.

What seemed important to us was that both Comings's work and the studies on mind-altering drugs were leading in the same direction, which gave us a starting point for our search—genes that affected monoamine signaling in the brain. Based on the combined evidence, we selected nine candidate genes to analyze. All of them had known genetic variations we could assay, and all affected monoamine signaling in the brain.

Reading DNA

Making DNA, as we've seen, is easy. Reading it is a lot tougher.

Imagine you have 35,000 books about the size of this one. Now suppose there's a single typographical error in one of the books. One wrong character in one word in one book out of an entire library. Your task is to find that one typo.

Difficult? You bet. Now suppose you have to check 1,000 other libraries, each also containing 35,000 volumes, to see whether or not they contain the same typo.

This was the task we faced in genotyping our nine candidate genes in 1,001 subjects. Most of the variations were single-base changes: an A in place of a T, or a G where there normally is a C. Others were differences in gene length due to sequences that were repeated different numbers of times. Obviously we couldn't determine the complete DNA sequence of each of the genes we wanted to check. That would be like reading all the way through 1,000 libraries of 35,000 books each—it would take forever. We needed a shortcut.

That shortcut was provided by a method called the polymerase chain reaction, or PCR for short—a method that won its inventor, Kary Mullis, a Nobel Prize. PCR allows scientists to copy, amplify, and analyze just one small part of the total genome. It works by mixing a person's total DNA together with two short pieces of synthetic DNA, called primers, that bracket the region of interest. The mixture is heated and cooled to allow the primers to bind to the specific region of total DNA that they correspond to; then an enzyme that can copy DNA is added. Within minutes, the original one copy of the target sequence has been multiplied to two copies. Now the heating and cooling cycle is repeated so that the two copies become four, and again to make eight copies, and so on and so forth until there is virtually nothing in the test tube but the gene of interest. The reaction is performed using a heat-stable enzyme isolated from a bacterium that grows in hot sulfur springs. That way it's possible to perform all the reactions in a single tube without needing to add fresh enzyme.

PCR is like having a copying machine that can pick out just one word in a library of books and blow it up to a size that can be seen across the room. There's still the problem, however, of spell-checking. If there were just one volume that had to be checked, it could easily be done by inspection. But examining nine different words, each 1,001 times, would be a challenge.

The same is true for DNA. We could take a single PCR-amplified gene from one person and determine its complete DNA sequence to check for changes. But that wouldn't be practical for inspecting nine genes in 1,001 people. Another shortcut was required.

The solution was provided by Max Myakeshiv, a recent Russian emigrant with a knack for computers and automation. Max realized that it is possible to distinguish single base changes in DNA by carrying out the PCR with two different primers: one corresponding exactly to the first version of the gene, the other corresponding to the second. If the primers were designed just right, and the reaction conditions were sufficiently stringent, the first primer would work only on the first version of the gene and the second primer would work only on the second version of the gene. This method, which was initially developed a decade ago, is called allele-specific amplification. (Alleles are different versions of the same gene.)

Max took the method one crucial step further. He made each of the primers with a special sequence that would bind to yet another primer, this one containing a fluorescent base in a hairpin structure. Hairpins are special types of DNA sequences that can form partially double-stranded structures because they are palindromes; in other words, they read the same forward as backward. "Madam I'm Adam" is an English palindrome. "AGGCTAGCCT" is a DNA palindrome. (In the case of DNA, the forward and backward reads must be done on the complementary strands; remember that an A base must be matched with a T base and a G base must be matched with a C base.) Max's hairpin primers also contained a fluorescent base capable of emitting light when excited with a laser beam, together with a quencher that absorbs that light if it is located close by. When the primer was in its beginning hairpin configuration, it didn't emit light because the fluorescent base and the quencher were physically close. When the primer was incorporated into a linear DNA molecule by PCR, the fluorescent base and the quencher were separated, leading to a brilliant fluorescence.

Max had one last trick up his sleeve. He made one primer with a red fluorescent base and the other with a green fluorescent base. Now all he had to do was amplify total human DNA with his two allele-specific primers and their corresponding fluorescent primers and . . . voilà! DNA containing one version of the gene glowed green, DNA containing the other version of the gene glowed red, and DNA containing both varieties glowed orange. It was a cute trick.

With Max's method there was no need to sequence. No need to use special enzymes. No need for expensive, time-consuming analytical methods. The entire reaction could be performed in one well of a microtiter plate, which is like a miniature muffin tin with ninety-six tiny indentations, each containing a separate sample of DNA. When the reaction was complete, the plate could be photographed or read by a fluorescent scanner. Reading DNA became as simple as reading a stoplight: red, green, or orange.

The Gene

Once Max's method was up and running, it didn't take long to genotype our 1,001 samples. The actual bench work was done by Stella Hu and Louise McHugh, two veteran technicians with tremendous skill, patience, and know-how. They have what scientists call "good hands."

Day after day, Stella and Louise labored over the miniature muffin pans containing the subjects' DNA samples. They painstakingly filled the wells with the reagents for the copying reaction, placed them in the thermocycler for PCR, then transferred them to the fluorescent plate reader. Each new set of primers had to be checked for specificity; each new data set had to be proofread for accuracy. Slowly but surely, the gene data accumulated in the computer, where it would later be linked to the behavioral and personality test score results. The reason for doing all this work was to understand more about cigarette smoking and mental conditions, not about spirituality, but exactly the same DNA data could be used for all three purposes.

It actually took much longer to analyze the data than it did to do the experiments. When I talked to my colleagues at the NIH about self-transcendence, they looked at me askance. My boss suggested that I focus on more tractable problems, and her boss admonished me to stick to cancer research. Even the people in my own laboratory—the bench workers who performed the gene analyses, the psychologist who administered the questionnaires, and the statistician responsible for putting the information together—were skeptical. The whole question seemed too far out for normal scientific analysis. We published many papers on cigarette smoking and personality and other mainstream topics, but the data on self-transcendence just sat there, unused, until I decided to take a look at it on my own time.

Once I finally did put the gene results together with the self-transcendence data, the results were fascinating.

The first gene on the list of candidates was D4DR, which codes for a receptor that senses the presence of dopamine—one of the monoamines—in the brain. It was a prime suspect for several reasons. In the Comings study, dopamine was the neurochemical most strongly associated with self-transcendence of all those examined. Comings speculated that this was because the D4DR gene contains an extremely variable repeated DNA sequence that changes the number of amino acids in the protein, which in turn alters the way it works in the brain. Some people have only three copies of this odd sequence; others have as many as eleven copies. D4DR's high association with self-transcendence might also be because it is expressed both in the limbic system of the brain—the seat of the emotions—and in the prefrontal region—the thinking part of the brain. Moreover, this gene had previously been linked to novelty seeking, a personality trait that is slightly correlated with self-transcendence.

Despite our high expectations, however, the data was negative. There was no association between the D4DR gene and any aspect of self-transcendence, no matter how the data was analyzed.

The next gene on the list was the serotonin transporter, which modulates the brain's supply of another important monoamine. Although Comings found only a weak link between this gene and self-transcendence, it was of interest to me because of the important role serotonin plays in emotions—especially negative emotions like depression, anxiety, and hostility. Scientists still don't know whether it's high or low levels of serotonin that are associated with negative emotions, because what they measure is the amount of serotonin circulating at one time, not the total amount of serotonin in the brain. What they do know, however, is that variations in serotonin levels are associated with negative emotions. But once again we found no association. It didn't make any difference whether people had the form of the serotonin transporter gene associated with feeling good or feeling bad—the self-transcendence scores of the individuals in our study were the same.

The next six monoamine gene variants we assayed were all the same. There was no association. It was beginning to look as if there was no such thing as a gene that influenced spirituality.

Then I met George Uhl, a scientist at the National Institute of Drug Abuse, who was speaking with me at a conference about addiction. I talked about cigarette smoking, while George focused on "garbage heads"—people who will use any drug they can get their hands on to get high. During a coffee break, George mentioned that he had found some new variants in a gene called VMAT2. The polymorphisms looked interesting, but he wasn't sure what they did. He asked if I would like to collaborate.

I'd never even heard of VMAT2, but my ears perked up when George explained that it made a protein that packaged all of the different monoamines into secretory vehicles—the biological packages that the brain uses to store its signal molecules. We'd been looking at each monoamine separately, one gene at a time—one gene for the dopamine D4 receptor, another for the serotonin transporter, and so on down the line. It was tedious work. The idea of examining a single gene that handled all the monoamines simultaneously was appealing. Not only would this gene be easier to analyze, it was also interesting because it is responsible for actions of the monoamines working together.

The next day, George e-mailed me the gene sequences with a map of every base that differed from one person to the next. We decided to concentrate on just one of these genetic variations, or polymorphisms: A33050C, a single base that could be either an A or a C. The name of the polymorphism reflects its exact nucleotide position on the human genome sequence of chromosome 10, where the gene is located. The reason for choosing this particular variation was that George's laboratory had shown that it was tightly linked to most of the other mutations present in the VMAT2 gene. This meant that we wouldn't have to assay every base change in the gene. This one alone would serve as a marker for all the rest.

Within a month, Max had worked out a color-coded fluorescence assay for A33050C. It didn't take long to genotype the entire sample and put the data in the computer, comparing the genotypes to personality test scores.

We hit pay dirt.

There was a clear association between the VMAT2 polymorphism and self-transcendence. Individuals with a C in their DNA—on either one chromosome or both—scored significantly higher than those with an A. The effect was greatest on the overall self-transcendence scale and was also significant for the self-forgetfulness subscale. With transpersonal identification and mysticism, the effect was in the same direction but just short of statistical significance. Somehow, this single-base change was affecting every facet of self-transcendence, from loving nature to loving God, from feeling at one with the universe to being willing to sacrifice for its improvement.

The VMAT2 gene variant containing a C—or "spiritual allele," as I began to think of it—was present on only 28 percent of chromosomes, compared with 72 percent carrying an A. But because both the C/C and C/A genotypes had increased self-transcendence scores, compared to the A/A genotype, it worked out that 47 percent of people in our study were in the higher spirituality group, as compared to 53 percent in the lower group—virtually half, which was exactly what we were looking for. While this one gene might not make one a saint, a prophet, or a seer, it was enough to tip the spiritual scales and predispose one toward spirituality.

The VMAT2 result made sense. We had predicted that spirituality would be related to monoamines, and now we had found a gene that was involved in how the brain uses these molecules. But how could we be sure that there was really a direct connection between the A33050C polymorphism and self-transcendence, and not just some indirect effect? Could God be tricking us?

The first thing we checked was the effect of gender on the association. As I mentioned earlier, females, on average, are more self-transcendent than men. Could that difference somehow account for the effect of the VMAT2 gene? To find out, we used a statistical method called multiple analysis of variance, or MANOVA, that determines whether two or more different variables effect an outcome independently or interactively. Our analysis confirmed the importance of gender for self-transcendence but failed to show any interaction with genotype. In other words, regardless of whether a person was male or female; the power of the gene was the same.

Another example of MANOVA would be to look at the effects of living in North Carolina and cigarette smoking on lung cancer. Both variables increase the chance of lung cancer, but North Carolina also has a higher smoking rate. Therefore, the variables work interactively. Living in North Carolina is not an independent effect.

Next we checked whether or not age played a role in the association between self-transcendence and the gene. Although age per se is not related to self-transcendence, it might have had an effect on the gene. But again, that turned out not to be the case. The relationship between VMAT2 and self-transcendence was the same in every age bracket, from college freshmen to octogenarians.

Race and ethnicity also were potentially important. Although most of our subjects were of European ancestry, there were enough African Americans, Asian/Pacific Islanders, and Hispanics in the study to check to see if there was a substantial difference between ethnic groups in the strength of the gene. There was not. If anything, the genetic effect was a tad stronger in the minority groups than in the white majority, albeit not significantly so.

The last check, and in many ways the most important, was to compare siblings. Again, siblings have the same racial-ethnic background and the same socioeconomic status. They go to the same schools and the same churches. They even have the same parental and grandparental influences. We could use siblings to control for variables we couldn't even imagine, much less measure.

The heart of the analysis was to compare pairs of brothers and sisters who had different VMAT2 genotypes. If our theory was correct, most males with the C variation in the gene would score higher than their brothers who had an A; the same would be true of sisters. Pairs where both siblings had either an A or a C were not included, since they could be compared only to other pairs, rather than to each other.

Only 106 pairs of our original 1,001 siblings had different VMAT2 genotypes. (This relative paucity of genetically discordant pairs was expected, since siblings are genetically correlated.) Of those, 55—a just-significant majority—fit the expected pattern in which the sibling with a C scored higher than the sibling with an A on the self-transcendence scale, 45 pairs had the opposite pattern, and 6 were ties. The result became more impressive when we looked at the extent of the difference. The brothers and sisters with a C scored on average 1.5 test units higher than their siblings with an A, a result that was remarkably close to, and in fact slightly exceeded, the difference seen in the population as a whole. Although the numbers were small, they clearly were pointing in the same direction. If the original finding had been just an artifact due to racial or ethnic stratification, the effect of genotype should have completely disappeared in the sibling pairs.

One of the pairs with different VMAT2 gene sequences were Tim Mooreland, the George Mason University senior, and his older brother, Peter. Tim, the more self-transcendent of the two, had C on one of his chromosomes, whereas Peter had two A's. They fit the pattern perfectly. Although this certainly isn't the only reason that Tim loves nature and feels more connected to the universe than Peter, it probably does account for at least part of their differences.

A Lead, Not an Answer

When I discussed the VMAT2 finding with my scientific colleagues, I got a range of responses.

"A God gene?" snorted one of them contemptuously. "That's gotta be nonsense. Have you replicated it?"

Much as I hated to admit it, my colleague had a point. The replication history of behavioral genetics is abysmal. All too often, new gene findings can't be repeated in different populations and have to be recanted. This is to be expected, given the small effect of most of the genes and the large numbers of behaviors and genes that have been investigated. Still, it's not reassuring.

In the case of VMAT2, we had shown that no particular subset of the study population failed to show association, which was encouraging. But we couldn't perform a truly independent replication, since we had analyzed every subject whom we had genotyped and phenotyped, and there was no mandate for us to take this particular aspect of our research further. At this point, all we could do was to leave the task of replication up to other researchers. It shouldn't be difficult. Many research groups now use the TCI for psychological and psychiatric genetic studies, and Max's method makes it simple to perform the genotyping. And frankly, most people are more prone to believe a replication when it's performed by a completely different research group.

Another colleague who heard me talking about "the God gene" asked, with eyebrows raised high, "Do you mean there's just one?" I deserved her skepticism. What I meant to say, of course, was "a" God gene, not "the" God gene. It wouldn't make sense that a single gene was responsible for such a complex trait. Besides, the numbers didn't add up. The twin studies showed that 40 to 50 percent of self-transcendence is heritable. But our analysis of the VMAT2 polymorphism showed that it raises self-transcendence scores by only a single point, or 7 percent of the mean—less than 1 percent of total variance. That means that most of the inherited effects on self-transcendence can't be explained by VMAT2. There might be another 50 genes or more of similar strength.

Other critics—scientists are, by nature, a skeptical lot—convinced me that I needed to analyze the relationship between the VMAT2 gene and all the other personality measures we had collected, to make sure the gene was influencing spirituality directly rather than through some other trait. To do so, we reran the statistical analysis using the different personality scales, rather than self-transcendence. We found that VMAT2 showed no correlation either to neuroticism or to intelligence. There also was no correlation to novelty seeking, harm avoidance, self-directedness, cooperativeness, extroversion, introversion, openness, agreeableness, or conscientiousness. The only personality trait that showed a significant association to the VMAT2 gene was persistence, a measure of "stick-to-itiveness." But the association was too weak to account for the gene's effect on self-transcendence. Further tests showed that VMAT2 did not influence spirituality via an effect on persistence. The two pathways were separate, like two rivers flowing in opposite directions from the same spring, rather than like a tributary flowing into a larger stream.

Several of my associates were quite curious about our findings. "That's cool," said one. "But how does it [VMAT2] work?"

It was a good question. It made me realize that finding a particular gene sequence associated with self-transcendence was not an answer to the question of where spirituality comes from—it was a lead. The God gene may be steering us in the direction of monoamines, but that was the beginning of the path toward spirituality, not the end. Where this path would take us I did not know. But I was eager to find out.

Five

Monoamines and Mysticism

We see that the mind is at every stage a theatre
of simultaneous possibilities.
—*William James*

It is Good Friday, April 20, 1962. Michael Young is twenty-three years old, a first-year divinity student at Andover Newton Theological School. He is sitting in a basement lounge at Marsh Chapel on the Boston University Campus. Around him are nineteen other divinity students, ten research assistants, a graduate student in the Religion and Society Program at Harvard University named Walter Pahnke, and a Harvard professor soon to gain considerable notoriety—Timothy Leary. Pahnke hands a clear gel capsule to Young, who swallows it.

The students are ushered into a small chapel equipped with two loudspeakers. The room soon is filled by the stentorian voice of Rev. Howard Thurman, the black chaplain of Boston University and mentor of Martin Luther King Jr. His Good Friday service is a moving mixture of music, scripture readings, and poetry about the life and crucifixion of Christ.

The capsule dissolves in Young's stomach, and some of its contents make their way to his brain. Strange new molecules waft across the convoluted pleats and folds of the cortex, the bights and bays of the hypothalamus and amygdala, the myriad streams and tributaries of the thalamus.

Young has never before ingested the chemical contained in the capsule, nor will he ever again. Yet it seems oddly familiar to his brain. There is something reassuring about its chemical appearance—the frizz of electrons over an aromatic ring like a mop of wavy hair, the tip of nitrogen surrounded by carbons like a tongue in a smiling mouth—that forms an effective disguise. As the drug bounces from one cell to the next, it passes by thousands of different molecules, each with its own distinct physiognomy. Most of these encounters are brief, like strangers passing on a crowded street. But every once in a while there is a flash of false recognition. The molecules stop, shake hands, then embrace. Bonds are formed, then broken.

The chemical changes in Young's brain are subtle, but the results are not. As perception changes, things start to look different. Colors are indescribably intense, shapes are bizarrely distorted, objects are surrounded by geometric figures. The world looks like a Jimi Hendrix poster—a twirling, surging, streaming sea of colors and shapes.

Young is hallucinating. As he will tell a reporter thirty-two years later, "I didn't know which was the real world. I couldn't keep straight what was happening inside my head and what was happening outside."

Young has a vision. He is in the center of a wheel of radiating colors. Each one represents a different path that he can follow. He must choose one or die, but he can't. He is paralyzed with indecision. He feels like his guts are being clawed out. Just as Reverend Thurman reads from the Edna St. Vincent Millay poem "I shall die, but that is all I shall do for death," Young lets go. He abandons his ego, transcends himself. He later recalls, "I had to die in order to live in freedom."

Hours pass. The hallucinations fade, and perception returns to normal.

Today Young is a minister who counsels addicts and warns his own children about the dangers of drug use. Yet the memory of his vision on Good Friday 1962 lives on. It was an important contributor to the deep sense of spirituality that he maintains to this day. Young summarizes the importance of his experience with these words: "Religious ideas that were interesting intellectually before took on a whole different dimension. Now they were connected to something much deeper than belief and theory."

The Good Friday Experiment

Michael Young was a participant in the largest and most systematic experiment that has ever been conducted on the potential of drugs to facilitate mystical experiences and spirituality. The drug that he consumed was psilocybin, a hallucinogen derived from mushrooms.

Psychoactive plants have been used for religious purposes for centuries. The peyote cactus, whose main active ingredient is mescaline, was used by the Aztecs before the birth of Christ. Mushroom-shaped stone artifacts from 100 B.C. to A.D. 300 have been discovered at multiple sites in Guatemala. Siberian shamans used the mushroom *Amanita muscaria* to induce religious trances. To this day, Native Americans in Mexico and the southwestern United States use mushrooms, morning glory seeds, and cactus in their religious ceremonies.

Drug-induced visions are, of course, not the same as spontaneous mystical experiences. Nevertheless, they may provide useful insights into some of the basic brain mechanisms underlying altered states of consciousness and their relationship to spirituality. As William James commented:

> As regards the origin of the Greek gods, we need not at present seek an opinion. But the whole array of our instances leads to a conclusion something like this: It is as if there were in the human consciousness a *sense of reality, a feeling of objective presence, a perception* of what we may call "something there," more deep and more general than any of the special and particular "senses" by which the current psychology supposes existent realities to be originally revealed . . . The most curious proofs of the existence of such an undifferentiated sense of reality as this are found in experiences of hallucination.

In terms of my research, the key point about Michael Young's hallucination is that it was caused by psilocybin, which is a chemical mimic of the monoamine serotonin.

Pahnke, who was both a physician and a theologian, designed the Good Friday experiment to explore the differences and similarities between drug-induced and spontaneous mystical experiences. He hypothesized that psilocybin would facilitate mystical experiences in religiously inclined individuals, and that such experiences would have long-term positive effects on behavior and attitudes. To increase the chances of success, the drug was administered to a group of like-minded believers participating in a communal religious ceremony—circumstances similar to those of the Indian tribes that use psilocybin for religious purposes.

Given the controversial nature of drug use and the subjectivity of mystical experiences, Pahnke realized that it was essential to apply a rigorously scientific approach. He chose a randomized, controlled, double-blinded, matched-pair design with an active placebo. In other words, the 20 subjects were organized into 10 matched pairs based on their religious background, experience, and personality. On the morning of the experiment, a coin toss determined who would get the psilocybin and who would get placebo. Nicotinic acid, which causes flushing and tingling of the skin, was used as the placebo to keep both the subjects and experimenters blind to who received the drug for as long as possible. Of course, once the hallucinations started it became obvious who got what; there's no way to fake a psilocybin trip.

Pahnke did everything possible to standardize the experiment. All the subjects were young white males. All of them attended the same theological school. All of them went through identical screening and briefing procedures and listened to the same religious service in the basement chapel, with its pews, altar, and religious symbols. All of them were interviewed and completed the same questionnaire within a week of Good Friday, and again six months later. The only difference was the chemical they swallowed at the start of the experiment.

In quantifying the effect of psilocybin on mystical experiences, Pahnke faced the same problem that Cloninger had when he wanted to measure spirituality: There was no obvious yardstick at hand. So he invented his own. Based on the writings of saints, seers, yogis, and scholars, Pahnke constructed a topology of mystical experience composed of nine components: unity, transcendence of time and space, positive mood, sacredness, sense of objective reality, paradoxicality, ineffability, transiency, and persistent positive changes in attitude and behavior. Just as Cloninger endeavored to measure spirituality independent of particular religious beliefs, so Pahnke based his scale on the supposition that "in the mystical experience there are certain fundamental characteristics that are universal and not restricted to any particular religion or culture."

I learned about Pahnke's work after discovering the apparent link between the VMAT2 gene, monoamine, and spirituality. But as I read through his mysticism scale, I was struck by the parallels to Cloninger's measure of self-transcendence. Many of the concepts are virtually identical. Pahnke's "transcendence of time and space" is nearly synonymous with Cloninger's "self-forgetfulness." Both of them refer to getting so wrapped up in something that you lose track of time and location. Pahnke's category of "unity," which involves a loss of normal sense perceptions and the distinction between self and other people and things, paralleled Cloninger's "transpersonal identification." The essential idea is to get beyond the confines of "me" versus "them."

The "sacredness," "sense of objective reality," and "positive mood" categories of Pahnke's mysticism topology paralleled Cloninger's "mysticism" subscale: how the experience feels real (even if paradoxical), feels special, and feels good. Pahnke's "ineffability," meaning indescribability, also has its parallel in the self-transcendence questionnaire, which asks, "Do you sometimes feel a spiritual connection to other people that can't be explained in words?"

Where Pahnke's scale differs from that of Cloninger is in duration. Mystical experiences are transient, whereas self-transcendence is long-lasting. This is because mystical experiences, by definition, are extraordinary. They are different from normal, everyday existence. Self-transcendence, by contrast, is part of a person's character. It's expressed every day of one's life. Although it may wax or wane depending on circumstances, it's always there to some extent. This is why Pahnke's final category, "persistent positive changes in behavior or attitude," has no parallel in self-transcendence, which is present in a person constantly.

In order to quantitate each subject's extent of mystical experience, Pahnke gave the interview tapes and questionnaire results to independent judges, who were blind to who took the drug and who received placebo. The results were calculated as the percentage of the maximum possible score, then averaged for the one-week and six-month assessments to give a final rating.

The results of Pahnke's experiment were striking. In every single category, the subjects who took psilocybin scored significantly higher than did the placebo controls. The young men who took the psychedelic drug had average scores of 64 percent of maximum, whereas classmates who received a nonpsychoactive placebo came in at the 14 percent level. It was clear from the results that taking a drug made the young theology students feel more at one with the universe and the people in it—more blessed, more sacred—as they listened to the pealing oration of Reverend Thurman.

Pahnke felt that for a mystical experience to be complete, a subject should score at least 60 to 70 percent in every category. According to this arbitrary cutoff point, four out of the 10 psilocybin subjects had complete mystical experiences; none of the controls did. Moreover, eight out of the 10 subjects who took the drug experienced completeness in at least seven out of the nine categories. Every one of the participants who took psilocybin had more extensive mystical experiences than did his matched control partner.

An Abiding Effect

The effects of psilocybin were powerful, but were they long-lasting? Would the mystical effects of the drug fade as the theology students completed their studies and moved on in life, or would they have a more durable effect?

Walter Pahnke never had a chance to find out; he died in a scuba accident in 1971. But his experiment lived on when Rick Doblin, a psychology student at the New College in Sarasota, decided to perform a 25-year follow-up. Working together with Mike Young, who by then was the minister of a Unitarian church in Tampa, he was able to locate and speak with 16 of the original participants in the study—nine from the control and seven from the experimental group. (One subject had died, one could not be identified, and two declined to participate because of privacy concerns.) Each of the subjects underwent a structured interview and filled out the same hundred-item mystical experiences questionnaire they last completed two and a half decades earlier.

Despite the passing of the years, everyone whom Doblin interviewed remembered their Good Friday experience. Their perceptions had changed remarkably little. For the seven subjects who took psilocybin, the average mysticism scores were 65 percent, virtually identical to the 64 percent scored 25 years earlier. For the nine control subjects, the average score remained considerably lower at 13 percent, little altered from the 14 percent level achieved in 1962.

Even more striking than the statistics were the comments recorded in Doblin's interviews. Most of the control subjects had only a dim recollection of the Good Friday experiment. For them it was just another church service. (The most notable effect was that several of them were inspired to try out psychedelic drugs later on.) By contrast, every one of the experimental subjects felt that their Good Friday drug trip had elements of a bona fide mystical experience and was a highlight of their spiritual life. They remembered Reverend Thurman's service in vivid detail and felt that their reaction to it had changed their lives for the better.

In sifting through the individual statements, it is remarkable how many of the recollections sound like descriptions of Maslow's "peak experiences," which are characterized by a feeling of wholeness and unity with the universe and everybody in it. For example, one subject had the following memory of his experience:

> It was a feeling of being . . . lifted out of your present state. I just stopped worrying about time and all that kind of stuff . . . there was one universal man, personhood, whatever you want to call it . . . a lot of connectedness with everybody and everything.

This sort of connectedness to the universe and everything in it is the hallmark of the transpersonal identification subscale of self-transcendence. Another subject recalled a similar sensation during his drug experience:

> The inner awareness and feelings I had during the drug experience were the dropping away of the external world and those relationships, and then the sudden sense of singleness, oneness.

Many of the subjects also experienced a distorted sense of time and space, a characteristic common to both Pahnke's topology of mysticism and the self-forgetfulness subscale of self-transcendence. Here's how one subject recalled his experience:

> All of a sudden I felt sort of drawn out into infinity . . . I felt that I was caught up in the vastness of creation . . . huge as the mystics say . . . I did experience this classic kind of blending . . . The main thing about it was a sense of timelessness.

Ineffability, another common thread in mystical experiences and self-transcendence, also was evident. As one subject remarked:

> I remember feeling at the time that I was . . . incapable of describing it. Words are a familiar environment for me and I

> usually can think of them, but I didn't find any for this. And I haven't yet.

Despite the fact that the students were not told whether they were in the experimental group or the control group, everyone who took psilocybin knew it. And yet, as is typical of mystical experiences, their hallucinatory experiences felt as if they were real. When Mike Young visited the men's room in the basement of Marsh Chapel, he imagined that some cigarette ashes in the urinal were black pearls. When he heard cars through the open window, he couldn't tell if they were real or an illusion. He no longer knew what was happening inside his head and what was taking place in the outside world.

The effects of the psilocybin, by and large, were positive. Many of the subjects commented on how the drug deepened their sense of joy, their openness to emotions, and their willingness to try new things. For others, the experience led to increased tolerance of others, appreciation of nature and the environment, and involvement in social and political causes.

One of the questions in the transpersonal identification subscale of self-transcendence is "Would you risk your own life to make the world a better place?" One of the subjects—who, like Mike Young, hallucinated his own death—felt that the psilocybin experience had made him more likely to answer yes. He recalled:

> When you get a clear vision of what [death] is and have sort of been there, and have left the self, left the body, you know, self leaving the body, or soul leaving the body, or whatever you want to call it, you would also know that marching in the Civil Rights Movement or against the Vietnam War in Washington [is less fearful].

Interestingly, even though five out of the eight experimental subjects were still working in the church when they were interviewed by Pahnke, their recollections are largely devoid of specifically religious imagery. One of the psilocybin subjects commented about his experience as follows:

> I don't think Christ or other religious images that I can remember came into it. That's the only reason I didn't think it was religious. I don't remember any religious images . . .

On the surface, it appears that a small capsule of psilocybin accomplished what Maslow, Cloninger, and many other psychologists, theologians, and philosophers struggled so mightily to achieve. It separated spirituality—or at least a perception of a spiritual event—from religion.

Because the chemical action in psilocybin is a mimic of the monoamine serotonin, the Good Friday experiment shows that altering monoamine signaling can profoundly alter consciousness. As such, it supports the connection between altered consciousness and spirituality.

But a drug trip, no matter how profound its effects, is not the same as a mystical experience. It's an artificial substitute—a spiritual saccharin. To delve further into the biology of spirituality, we need to understand more about how monoamines work normally in the brain to produce the greatest hat trick of biology: consciousness.

Six

The Way Things Seem

One must say truly, I think, that personal religious experience has its root and center in mythical states of consciousness.

—*William James*

Our brains are bombarded with information every second of every minute of the day, day in and day out. There is the data we receive from the outside world through our senses: sights, sounds, smells, tastes, and touches. And then there is the news that one part of the brain receives from other parts of the brain: memories, feelings, thoughts, and dreams.

The sheer magnitude of the data is staggering. As I write this, I am on a jet flying from my father's house on Long Island to my home in Washington, D.C., working on my laptop and listening to the Rachmaninov Vespers through a headset. If I were to take a snapshot of this moment, it would show a blue computer screen filled with words, a gray keyboard resting on a beige tray top against a background of green seat upholstery, and a black window with flickering lights in the distance. If I transferred that simple picture to my computer as a digital image, it could easily consume about 100,000 bytes of information. Were I to simultaneously record one second of the Vespers as an audio file on my computer, it could occupy about 250,000 bytes of information. Now suppose I made a snapshot and audio recording ten times a second for the entire sixty-minute flight. By the time I touched down, I would have accumulated more than 10,000,000,000 bytes of information. Without a compression program, the hard drive on my laptop would overflow before we even reached Delaware.

Now imagine that I tried to record all the thoughts that passed through my mind during the flight: the unfamiliar stab of anxiety as we took off, the bewildered grief as we passed by the gaping hole at Ground Zero in the skyline of Manhattan, and then the nagging question of where I parked the car when I left Washington Friday evening. How many bytes would be required to record all these memories remembered, all these feelings felt? Certainly as many as were needed for the physically sensed sights and sounds.

Now multiply these numbers by all the seconds of a lifetime, and the amount of data managed by the mind becomes truly staggering. It would seem more than any supercomputer could handle.

My brain is certainly no computer. It can't even beat my puny laptop at chess. And yet it receives, sorts, processes, and analyzes this incredible volume of data with the greatest of ease to produce the most remarkable of all products: a seemingly coherent picture of the world that surrounds me.

This is consciousness—our awareness of our surroundings and ourselves. It is at once both the most commonplace and the most mysterious of all life processes.

Nobody knows exactly how consciousness works. There are plenty of theories—one of which I present below—but the detailed mechanism remains obscure. Nevertheless, there are several features of consciousness that are universally recognized. Most of these were described by William James more than a century ago.

It is important to describe these general properties of consciousness, because they are what change during mystical experiences.

First, consciousness is selective. As I write this, the only things of which I am fully aware are the words in front of me on the computer screen and the music coming through my headset. I know there is a tabletop under the computer and a window to my left, and that the people in the seat behind me are having a conversation about their weekend in the country—and yet none of these facts is truly "on my mind." They do not occupy my attention. Although our brains are capable of processing a vast quantity of information, they can do so effectively only one topic at a time.

Second, consciousness is continuous. Were I to close my eyes, take off my Walkman, turn around in my seat, and then reopen my eyes, I would be presented with a quite different set of sights and sounds than I had encountered a few seconds previously. Nevertheless, I would know full well that I was still in the same place and that my surroundings had not changed at all. It would be a seamless transition, because consciousness has a built-in mechanism that connects each moment to the one before and the one after. Consciousness can fill in the gaps between "data points."

The third key feature of consciousness is that it is personal. It is yours and yours alone. You can describe to other people what you see or how you feel, but the sights and feelings themselves are not transferable. This is why you always experience the world from your own viewpoint, either as a direct participant or as an onlooker, not from the perspective of somebody else. That is also why the one thing that you can always be certain about is that you are you and not somebody else. Imagine a world in which this was not the case—in which you confused your own identity with that of others. Confusion would reign.

William James made one other observation that remains central to our understanding of consciousness: It is a process, not a thing. As such, the key question is not what a particular state of consciousness consists of, but how it arises.

Most current theories divide the process into two stages. The first is core consciousness, also known as primary consciousness. This is what lets us interpret raw data as integrated scenes. Core consciousness is what makes me see a computer with words in front of me, rather than a jumble of millions and millions of unconnected pixels of color. The second stage is known as higher, or secondary, consciousness. This is what incorporates the autobiographical me into my perception of the world around me. It involves the ability to recognize one's own acts and feelings, to conceive of past and future as well as present, and to be cognizant of purely mental images. It is higher consciousness that reminds me that I am the one who wrote the words on the screen, reminds me of the reason why, and what I am supposed to do with them next.

Core consciousness is required for higher consciousness, since you couldn't know who you are if you didn't have awareness. But higher consciousness is not a requirement of core consciousness. Core consciousness appears to be present, to one degree or another, in many higher life-forms. Lions need it to hunt; dogs need it to fetch. Higher consciousness, on the other hand, seems to be uniquely human, or at least limited to us and our close cousins, the primates. Since higher consciousness allows us to manipulate purely abstract concepts and see ourselves in relation to them, it more closely resembles spirituality than does core consciousness.

Mind or Matter?

One aspect of consciousness has puzzled philosophers, theologians, metaphysicists, psychologists, and biologists for centuries and probably will continue to do so for many more. Sometimes called the mind-body problem, it boils down to this: Can consciousness be explained scientifically?

By a scientific explanation, I mean one that can be expressed in terms of the basic principles of chemistry and physics. Proponents of this view often are called "materialists" because they believe that all mental processes can ultimately be accounted for by a few basic physical laws. Most scientists, including myself, are materialists.

It is important to emphasize at the outset that a scientific explanation of consciousness need not be expressed only in terms of basic physical and chemical properties of atoms and their constituents. Higher levels of organization are undoubtedly involved, even if not yet fully understood. Indeed, a strictly atomic explanation of consciousness may never be achieved. Even if it were, such an explanation might be too finely grained and complicated to be comprehended. Rather, the explanation must be fully consistent with the fundamental laws of the natural sciences.

What level of explanation is most relevant for consciousness? The most likely answer involves groups of interacting cells—specifically, assemblies of neurons, the cells that carry information in the brain. A single neuron is not a particularly impressive or smart object. Basically, all it does is convert one type of chemical or electrical signal into another. But the intricate interactions between many such units acting together can generate truly amazing complexity. Finer levels of organization—such as the atomic structure of the receptors and other macromolecules that neurons use to signal one another—also are important to consciousness. So, too, are grosser levels of brain organization, like the hypothalamus and the amygdala. It is at the level of cellular interactions, however, that a breakthrough in our scientific understanding of consciousness is most likely to be achieved.

Nonscientific explanations of consciousness invoke a variety of principles above and beyond natural law. Proponents of such theories are sometimes called "dualists," because they believe mental and physical processes are fundamentally different in kind. The oldest such explanations are religious: Consciousness is a gift from God. Seventeenth-century thinkers were particularly adept at this branch of philosophy. Nicolas Malebranche, for example, developed the concept of "occasionalism," which holds that God is the only true causal agent. According to this theory, looking at an apple doesn't cause you to see red, it's an occasion for God to make you see red. The most famous of these philosophers was René Descartes ("I think therefore I am"), who believed that the mind and body are distinct substances that communicate through the pineal gland, which he believed was the seat of the soul.

One need not invoke God to avow the dualist position. Modern philosophers have developed a number of alternative explanations of consciousness that maintain the distinction between mental and physical processes without resorting to faith in a higher power. Among these are anomalous monism, eliminative materialism, and token epiphenomenalism, and for most people they are as difficult to understand as they are to pronounce. Other thinkers simply proclaim that the origins of consciousness are too difficult to be solved with our present knowledge.

The mind-body debate has been going on since at least the time of the ancient Greeks. Aristotle believed that body and soul were two aspects of the same thing; Plato held that the soul was distinct from the body and could survive outside it. The issue will not be definitively decided until the actual mechanism of consciousness is understood. But there is already ample reason to believe that a scientific explanation of consciousness is at least possible. Let me offer two arguments, one quantitative and the other historical.

First the numbers. How can a bunch of brain cells store all the sights and sounds and memories and feelings and dreams and hopes and fears of a lifetime—and keep them in order, no less?

The answer, scientists suggest, is combinatorials.

The brain contains a total of about 100 billion neurons, written in scientific notation as 10^{11}. That's a comprehensible number; it's about the same as the United States' Medicare budget for this year, expressed in dollars.

Each neuron contains on average about 1,000 (10^{3}) branches that form connections, known as synapses, with other neurons. Multiplying the number of synapses by the number of neurons gives a total of 100 trillion synaptic connections (or $10^{11} \times 10^{3} = 10^{11+3} = 10^{14}$). That's still a reasonable number—about four times greater than the net worth of the entire United States in year 2001.

Each synapse can have any of a number of different functional states corresponding to various levels of electrochemical activation or suppression. This is the way that neurons tell one another what's going on. If we suppose that on average there are 10 possible states per synapse, which probably is an underestimate, how many different neural states are possible?

This is where combinatorials come in. To get the answer, we can't simply multiply 10×100 trillion, which would give a prosaic 10^{15}. Rather, because of the fact that when one neuron is in one state each other neuron can be in any of ten other states, we must raise 10 to the 100 trillionth power. That gives $10^{10\times14} = 10^{100000000000000}$.

This is very decidedly not a reasonable number. It is far more than all the money that has ever existed or ever will exist on this planet, even if counted in pennies or lira or centimes. It is more than every grain of sand on every beach, every star in every galaxy, every elementary particle in the universe—or all of them combined. If you started speaking this number at the beginning of time, you would only be partway through when time ended. It is a number that would make Carl Sagan blanch.

Even if only 0.01 percent of all these possible states actually meant something, that would still leave more than $10^{100000000000}$ meaningful configurations of the brain. That's plenty to receive and process all the events and feelings of a lifetime. The take-home message is that our brains have more than enough computing power to handle every possible situation, to process every conceivable thought.

Now let me give you the historical argument. Since the beginning of time, people have been faced with phenomena that they cannot understand with the tools and knowledge available to them. The sun rises and sets, people are born and die, grape juice becomes wine, and wine becomes vinegar. The three most common explanations are "God did it," "The devil did it," and "It can never be understood." Those answers almost always turn out to be wrong.

Consider the question of where life comes from. It was believed for more than twenty centuries that life could spontaneously arise from nonliving materials such as moldy hay and decaying fruit. Some people even believed it was possible to create mice by placing sweaty underwear and wheat husks together (one more reason not to eat crackers in bed). In the course of trying to help a vintner rid his wine of undesirable contaminant, Louis Pasteur performed a simple experiment to determine the origin of the guilty microorganisms. He took rich broth, sterilized it by boiling, then let it sit out in the air in a special flask with a curved swanlike neck that would trap any dust particles. Nothing grew. By contrast, similarly treated flasks without the special neck rapidly became cloudy with bugs. Pasteur concluded, "There is no known circumstance in which it can be confirmed that microscopic beings came into the world without germs, without parents similar to themselves."

We now face the same situation in regard to consciousness that Pasteur confronted in the middle of the nineteenth century with regard to the origins of life. We have catalogued some pertinent features of the phenomenon, identified a few of the key players, and proposed various mechanistic models. We still don't understand how it works. But that doesn't mean that it *can't* be understood, just that it's complicated. What we need is a new Pasteur.

Edelman's Brilliant Fire

Something definite happens when to a certain brain-state a certain "sciousness" corresponds.

—*William James*

Gerald Edelman would like to be the Pasteur of consciousness. He certainly has the qualifications: a Nobel Prize in biology, the directorship of the San Diego Neuroscience Institute, a probing intellect and ceaseless curiosity, and—like Pasteur—abiding faith in the experimental method. Most important, he has a theory of consciousness. A testable, scientific theory.

Edelman's theory, which is described in his books *Bright Air, Brilliant Fire* and *A Universe of Consciousness*, starts by distinguishing between two fundamentally different types of nervous system organization. The first is the thalamocortical system, which is named after its two principal structures, the thalamus and the cerebral cortex. The thalamus is a dense knot of cells buried deep within the center of the brain. Looked at in cross section, it appears like an egg lying on its side, surrounded by packing material. The thalamus acts as a relay station—the "last pit stop" for information flowing from the outside world to the thinking part of the brain. It receives inputs from the visual system, the auditory system, and the touch sensory system. The cerebellum and several limbic structures also send neuronal projections, which are called axons, to this relay center.

The thalamus sends its outputs to the cortex—the wrinkled, convoluted outer layer of the brain. The cortex is what you see when you observe a pickled brain in a glass jar. It is geographically divided by an intricate system of ridges and valleys into four pairs of lobes: frontal, temporal, parietal, and occipital. The cortex is an evolutionarily modern part of the brain and is responsible for many key functions, including memory, planning, the control of movement, and perception.

The second, more ancient type of nervous system organization is the limbic–brain stem system. The brain stem, which is a general term for the area between the spinal cord and thalamus, is the connector between the body and the brain. It includes a number of structures, including the medulla, pons, and reticular formation, which are involved in basic life processes like breathing and maintaining blood pressure. The limbic system includes the amygdala and hypothalamus, which play key roles in emotional responses, as well as the hippocampus, the brain's memory archives.

The thalamocortical and limbic–brain stem systems evolved to carry different types of information and use different communication styles. The purpose of the thalamocortical system is to receive a dense and rapid series of signals from the outside world through sight, sound, taste, smell, and touch and to coordinate motor movements, perception, and categorization in response. Its bits and pieces are designed for speed and pinpoint accuracy, with communications occurring in milliseconds to seconds. This is the system that allows a leopard to leap on its prey.

The limbic–brain stem system is built for comfort, not for speed. It lets the brain know what's going on in the body—down in the heart, the gut, and the bloodstream—so that the brain can make the necessary adjustments: turn up the pressure a bit, or turn down the heat. Communications are slow, on the range of seconds up to months, as befits such a large and important bureaucracy. This is the system that made the leopard hungry in the first place.

According to Edelman's theory, consciousness arises through the process of communication both within and between these two systems. Communication within the thalamocortical system involves a large, three-dimensional network of circuits that connect thousands of different groups of neurons. Almost all of the connections are reciprocal, allowing the constant back-and-forth sharing of data. Scenes are generated by the combination of many different inputs to this dynamic core from conceptually related stimuli.

To take a gross example, if we were talking face-to-face, my eyes would be watching your lips as my ears listened to your voice. But in my brain, the auditory center would "know" what the eyes were seeing, and the visual center would "know" what the ears were hearing and would automatically and effortlessly integrate the two inputs. That way, even if I missed hearing a few words because of an extraneous sound, I would be able to fill in the gap in the conversation. (The reason that it's so funny to watch dubbed films is the disconnect between lip motions and sounds.) Our brains constantly generate scenes by this sort of back-and-forth mapping, only on a much more rapid and massive scale between all the many different parts of the thalamocortical system.

The components of the limbic–brain stem system communicate with one another in a different way. The network looks like a fan rather than a mesh, with its base located in a limited number of cell groups (called nuclei) in the brain stem and hypothalamus. The type of information conveyed is about values, not sights or sounds—whether something feels good or feels bad, not whether it is red or green. The limbic–brain stem system is about emotions, not about scenes.

It is in the limbic–brain stem system that monoamines intersect with consciousness. The cells of the limbic–brain stem system communicate with the brain through these neurotransmitters, which can make a person either feel good or feel bad, depending on which chemical is released and what type of cell it encounters.

Most animals have thalamus, cortex, brain stem, and limbic systems, yet they lack our sophisticated brand of consciousness—and our spirituality. The difference may lie in the way our brain systems communicate back and forth with one another. According to Edelman's theory, the thalamocortical and limbic–brain stem systems were linked during evolution in such a way that their activities could complement each other. The brain matches perceptual categories generated by the thalamocortical system with value signals originating from the limbic–brain stem circuit. This category-value system then feeds back through the thalamocortical system for further categorization, creating a continuous loop of correlations between events and categories. Thus, human consciousness is, in Edelman's words, a kind of "remembered present." Not only do we see and smell and taste things, we know what they are and what they represent for us.

When a leopard devours its prey, it's just a hunk of meat. When you dig into a nice juicy steak, it's far more than that; it is intimately connected to your memory of the first time you had a backyard barbecue, your dread of the extra half hour you'll need to spend on the treadmill to work off the calories, and your worry about the reaction of your vegetarian daughter. What is unique about human consciousness is our ability to associate scenes and senses with emotions and values.

Mapping Consciousness

Edelman has more than just pretty words to support his theory. He has an experiment.

If you simultaneously look at two different images, one through each eye, you will be conscious of only one of them at a time. The perceptually dominant image will switch every few seconds. This is a consequence of the selective nature of consciousness that James noted more than a century ago.

This phenomenon of binocular rivalry, as it's called, can be used to map the brain areas that are involved in consciousness. The actual experiment consists of placing a large hair dryer–like device containing 148 sensitive magnetometer coils on the head of a subject sitting in front of a projection screen. By having the subject wear colored goggles, a red vertical grating flickering at one frequency is presented to the left eye and a blue horizontal grating flickering at a slightly different frequency is presented to the right eye (or vice versa). The subject signals which pattern he or she is aware of by pressing a switch with the right or left index finger. Meanwhile, the magnetometer is used to record the position and intensity of the electromagnetic activity generated by the brain.

The trick behind the experiment is to focus on the magnetic signals with the same frequency as the flickering light the subject is conscious of at that time. By subtracting the dominant from the nondominant signal, the areas specifically responsive to the red or blue light can be identified. Edelman's theory predicts that consciousness is not limited to any one particular region of the brain since it involves interactions between large populations of neurons located in many different regions both within and beyond the sensory apparatus.

That's exactly what was observed. The magnetoencephalograms showed signals over large portions of the cortex, including the occipital, frontal, and temporal lobes. Different subjects showed considerable variability in their patterns, but all of them were spread out among many different regions. The most important finding was that the signals extended well outside the visual cortex, the part of the brain responsible for receiving signals from the retina. Just as predicted by Edelman's theory, being conscious of something involves more than simply seeing it.

The Gift of Monoamines

So far so good for Edelman's theory. Consciousness links the physical senses to the emotions through widely distributed brain networks. But that still leaves the critical question of how those emotions are generated. What is it that actually makes us feel happy, sad, excited, or anxious?

This is where the monoamines and the VMAT2 gene come to the fore. The monoamines are the biochemical mediators of emotions and values. They are what make us feel. But monoamines are not freely available to the brain; like a gift, they first need to be wrapped up and unwrapped. That's where the VMAT2 gene plays a critical role.

The monoamines are so named because they contain a single amine group—a nitrogen atom surrounded by three hydrogen atoms—tethered to the molecule by a short chain of carbons. The monoamines fall into two classes. Dopamine, adrenaline, and noradrenaline are part of what are called catecholamines—an amine group attached to an aromatic ring with two protruding oxygen molecules. Serotonin, on the other hand, is an indole amine; here, the amine group is leashed to a double ring structure without oxygens.

All of the monoamines are small, about the size of an aspirin molecule, and are synthesized from amino acids that are found in common foods and proteins. That's why what you eat can affect how you feel. The reason you reach for a candy bar instead of a hot dog when you feel depressed isn't because candy is sweet; it's because sugar in the blood makes more serotonin in the brain, which relieves feelings of depression.

Monoamines are produced in neurons, the cells that carry information in the brain. The way that the brain "knows" what you are sensing and how you are feeling is through one neuron talking to another, often using the language of monoamines.

Although neurons come in many shapes and sizes, they all have certain structures in common: a cell body, dendrites, axons, and nerve terminals. The cell body is like the main root of a tuber—big and bulky. It carries out most of the maintenance operations of the cell. The dendrites, which resemble root hairs, short and spiky, are the cell's antennae—its way of gathering information. An axon, which looks more like a tree trunk, long and sturdy, carries information away from, rather than toward, the cell. Each axon branches many times, sending out hundreds or thousands of limbs that finally end in small knobs called terminal buttons. These buttons carry out the business of an axon, spitting out the chemicals that transmit signals to adjacent neurons. The tiny gap between the terminus of one nerve cell and the dendrite of another is the synapse.

The cells that produce monoamines have their bodies located deep within the middle of the brain but send their branches out far and wide. The system for serotonin, the brain chemical involved in negative emotions, is particularly widespread, projecting from the midbrain raphe nuclei out to the forebrain, neocortex, olfactory bulb, hippocampus, thalamus, cerebellum, and even down into the spinal cord. Dopamine, the brain's activator or "feel-good" chemical, is produced by two different groups of cells. One cluster of neurons is based in the ventral tegmental area and sends its axons mostly to the forebrain, the prefrontal cortex, the amygdala, and a special area called the nucleus accumbens. The second group of neurons have their home base in the substantia nigra and project mostly to the striatum, which is involved in the control of motion. These are the cells that, in Parkinson's disease, degenerate, resulting in a loss of bodily motor control and tremor. Finally, noradrenaline, which plays a role in stress and alertness, while made in just a few thousand cells in the locus ceruleus, sends out a dense network of projections to many different regions, including cortex, cerebellum, and hippocampus.

Although these are the cells where monoamines are produced, they must be gift-wrapped before they are sent on their way.

Cells wrap up monoamines with membranes—strong, flexible material made out of proteins, sugars, and fats. The membranes form tiny spheres called vesicles. A single nerve terminus may contain thousands of vesicles, some small and others large. They mostly are clustered around the very end of the terminal button in the area known as the release zone. Some of them actually touch the edge of the cell.

Wrapping up the monoamines is hard work. This is where the vesicular monoamine transporter encoded by VMAT2 enters the picture.

The VMAT2 transporter is a long, snakelike molecule that weaves in and out of the vesicular membrane. The head and tail of the VMAT2 are both securely tucked inside the vesicle. The body crosses the membrane twelve times to form six coils running from inside to outside and back again. At each juncture with the membrane there is a transmembrane helix, a tight spiral of amino acids that fits perfectly into its slippery, greasy coil.

The transporter forms a channel across the membrane that acts as the border crossing for monoamines. Entry into the security of the vesicle comes at a price. For every monoamine that is pumped in, two protons (positively charged hydrogen ions) must be removed.

Once the monoamines are packaged into vesicles, they sit there, completely shielded by the vesicular membrane, waiting for something to happen.

That "something" occurs when a hormone or another neurotransmitter stimulates the neuron that produced the monoamine, sending a brief pulse of electrochemical energy coursing along the length of the axon. Once it reaches the membrane at the terminal button, it opens a tiny channel that lets in calcium ions. Calcium does more than build strong bones; it also binds to proteins and changes their shape. When calcium binds to proteins at the juncture between a vesicle and the cell membrane, it causes them to move apart from one another, forming a small opening. Monoamines begin to leak out of the cell. It's like opening the gate in a canal lock just a tad. Once the flow has started, there's no stopping it. Soon the entire contents of the vesicle are dumped into the synapse. The gift has been unwrapped.

The monoamines don't stay in the synapse for long. They almost immediately begin to do what neurotransmitters are designed to do: They signal.

Monoamines convey information by interacting with receptors, protein molecules that sit on the surface of cells and communicate between the inside and outside worlds. First, the monoamine snuggles into a binding site—a nook of the molecule designed to perfectly complement its shape—located on the part of the protein outside of the cell. This slightly changes the conformation of the receptor, including the inside part, which in turn activates another type of protein called a G protein. The G protein breaks away from the complex and activates the synthesis of cyclic AMP, a small signaling molecule that influences many cellular processes. Cyclic AMP is known as a "second messenger," because it now carries the information that started out in the hands of a monoamine.

The monoamines use a fundamentally similar biochemical signaling process, yet their effects are quite diverse. The secret to this diversity is in the receptors.

First, there are different classes of receptors for different monoamines. The receptors for serotonin, for example, don't respond at all to dopamine or noradrenaline, while dopamine works well on its own receptors but doesn't do a thing for serotonin or noradrenaline receptors. This specificity derives from the exact fit between the binding nook of the receptor and the shape of the monoamine. Each monoamine has its own unique shape and size that fits into its particular binding site and no other. Putting serotonin into a dopamine receptor would be like dressing Aretha Franklin in an outfit designed for Cher. It doesn't fit.

Second, there are many different receptors, even for a single monoamine. In the case of serotonin, for example, more than a dozen distinct receptor proteins have been identified. Although all of them have a similar binding nook, their internal structures are quite different, leading to diverse effects on intracellular biochemistry and signaling. This is why serotonin might turn on signaling at one synapse but turn it off at another.

Last, different receptors are made at different places in the brain and on different types of cells. A receptor expressed in the hypothalamus may have an effect quite different from the same receptor in the amygdala. Even receptors on two different cells in the amygdala will probably not act the same way. In brain chemistry, as in real estate, location is everything.

An Essential Gene

Just how important is VMAT2 in monoamine signaling?

To find out, three different teams of scientists—one led by George Uhl at the National Institute of Drug Abuse, a second at Duke University, and a third a collaboration between researchers at the University of California San Francisco and at Columbia University—made "knock-out" mice.

Knock-outs are an effective way to figure out what a gene does. The method is technically complicated but conceptually simple. Basically, genetic engineering is used to create organisms that either completely lack the gene being studied or contain only one instead of the normal two copies. By comparing the strains with zero, one, or two copies of the relevant gene, it is possible to infer its function. If the gene controls a black fur pigment, for example, the mice will be either white, gray, or black, depending on whether they have zero, one, or two copies of the gene.

The scientists started their engineering feat by making a mutated form of the VMAT2 gene that was incapable of producing any enzyme. Next, they force-fed this "turned-off" gene to cultured embryonic stem cells and selected for those in which it replaced the normal gene. The final step was to graft some of the genetically altered cells into fresh embryos and transplant them into female mice. Embryonic stem cells have the remarkable capacity to regenerate into any tissue in the body. As a result, some of their babies had the mutated VMAT2 gene in their sperm or egg cells and could pass it on to the next generation. By breeding these mice to one another, it was possible to produce just what the scientists wanted: rodents with zero, one, or two VMAT2 genes in every cell in the body.

The results of the knock-out experiment were dramatic. The mice without the VMAT2 gene were runts. They grew to less than half the size of normal mice and showed little interest in suckling or eating. Even when, to reduce competition, their normal littermates were removed, the mutants did not eat much. They were inactive, spending most of their time lying or crouched on the cage floor. When the investigators marked the side of a knock-out mouse with a delible ink mark, it was still there the following day. This wasn't because the mice were paralyzed; when they were pinched or prodded, they jumped sharply. They just didn't seem to have the will or interest to move normally.

The most striking effect of the knock-out was premature death. About half of the knock-out baby mice died within three days, and virtually none made it to two weeks.

To find out why, the scientists looked at the brains of the knock-out mice. Anatomically, their brains were fine—about the same as the brains of normal mice, only a tad smaller. All the cells that normally produce monoamines were present and had the usual morphology. When the scientists measured the actual amounts of the brain monoamines, however, there was a striking difference. The knock-out mice had drastically reduced concentrations of serotonin, dopamine, and noradrenaline. The levels of all three critical monoamines were down by at least a hundredfold. They could barely be detected at all.

Why would deleting the VMAT2 gene reduce monoamine levels so dramatically? After all, the transporter is supposed to gift-wrap monoamines, not produce them. Subsequent experiments showed that the answer entailed degradation. Although the enzymes involved in synthesizing and degrading monoamines were present in the mutant mice in normal amounts, the breakdown products of the monoamines were greatly elevated. This means that the mice initially produced normal amounts of the monoamines, but because they were never wrapped up in vesicles, they were rapidly degraded by cellular enzymes. To prove this, the scientists treated some of the mice with an inhibitor of monoamine oxidase—one of the degradative enzymes—and showed that the serotonin levels were corrected. They also treated mice with amphetamines, which release dopamine from nonvesicular sources; this partially restored movement, feeding, and survival.

The knock-out experiment shows that the vesicular gift-wrapping of monoamines does more than make a pretty package to deliver to the receptors. It protects the contents from being degraded and lost. It also makes clear that VMAT2 is a vital gene—without it, animals die. But the experiment doesn't tell us much about the role of the gene in spirituality or even consciousness. It's hard to measure something when the experimental subject is dead. For this, other approaches were needed.

Sensitive People

Ari has been dancing for many hours. Whirling and swirling, stamping her feet and raising her arms in joyous exaltation, she feels connected to every other dancer in the room—no, in the entire universe. She rides the rhythmic music into a magical, mystical realm, overcome with gratitude and love. She feels close to God.

Ari is neither a whirling dervish nor a Kong tribeswoman nor a Native American ghost dancer. She is a typical American teenager, a college freshman who took a dose of Ecstasy before she went out dancing. Ecstasy, which has the forbidding chemical name of 3,4-methylenedioxymethamphetamine, changed the concentration in her brain of a monoamine—serotonin, to be precise.

I mention Ecstasy not because it evokes a genuine mystical or religious experience, but because it illustrates how even a single monoamine can influence aspects of consciousness. When Ecstasy hits the brain, it releases a flash flood of serotonin. This has two psychological effects. First, it increases sociability and intimacy. Suddenly, everyone seems like a lifelong friend. The user has no enemies, only potential lovers, and all the world is at peace. (If a complete stranger comes up to you in a dance club and starts stroking your face, he or she is probably "rolling" on Ecstasy.)

The second major effect of the serotonin released by Ecstasy is mood elevation. Feelings of depression, anxiety, and inadequacy are instantly erased. Dark-tinted glasses turn to rose. The glass that seemed half empty is suddenly half full. The user feels—well, ecstatic.

If this description makes Ecstasy sound so good that you want to go out and try some right away, hold on. It turns out that the positive effects are only transient; the negative ones, however, are permanent. The reason is that Ecstasy is a potent neurotoxin that kills the very cells that it caused to release serotonin. The long-term result is a depletion of brain serotonin that can lead to severe depression, loneliness, and even suicide.

Fortunately, there is a more long-lasting and less harmful version of Ecstasy that you may be familiar with: Prozac. This drug, and its chemical cousins such as Paxil and Zoloft, inhibit the action of the serotonin transporter, a membrane protein that reabsorbs excess serotonin from the synapse. Blocking this monoamine-specific transporter causes a change in serotonin levels similar to that produced by Ecstasy. It takes much longer, though, to have an effect—

weeks instead of minutes—and does so without killing any cells.

The effects of Prozac are remarkably similar to those of Ecstasy, albeit more subtle. Negative feelings are replaced by more positive ones, which is why the drug is commonly prescribed to people who feel depressed. The drug also makes people more sociable and cooperative. In one laboratory experiment, Prozac or a placebo was given to subjects who were then asked to solve a jigsaw puzzle together. The ones who took the drug were much more helpful to one another than those who did not. Prozac does all this without harming brain cells, which is why it is one of the most heavily prescribed drugs in the world.

There is a natural, genetic version of Ecstasy in humans. It results from a variation in a part of the serotonin transporter gene responsible for copying the DNA into messenger RNA. People who inherit one version of this polymorphism make more RNA, and therefore more functional transporter protein, than people who have the other gene.

The effects of this genetic variation are similar to those of Ecstasy and Prozac. Individuals with one form of the gene tend to be worried, anxious, sad, and emotional—not to mention shy, uncooperative, and suspicious. They may be overly sensitive. Folks with the other common version of the gene are more likely to be carefree, happy, and stable. They also are more apt to be your friend, since they are genetically predisposed to be sociable, extroverted, and trusting.

In other words, serotonin affects consciousness in many ways that are connected to self-transcendence and spirituality. It alters perceptions, one aspect of mystical experience. It increases sociability, an aspect of transpersonal identification. It elevates mood, an aspect of the perception of sacredness. All of this can be influenced by a single monoamine.

Feeling Good

Imagine that next to Ari on the dance floor is Tomas. His arms wave as he dances, his hips grind, his eyes bulge, sweat beads his brow. He is feeling good, flying high—the best dancer in the room, in his estimation. Tomas doesn't feel just feel close to God, he feels like he *is* God.

Like Ari, Tomas is under the influence of a monoaminergic drug, in this case cocaine. Cocaine releases dopamine, the brain's reward chemical. Although dopamine and serotonin are chemically similar, there are big differences in what they do. Dopamine makes people feel good rather than just not bad, sociable rather than just not hostile. If serotonin is the brain's stick, dopamine is its carrot.

The reinforcing properties of dopamine were discovered in 1954 by accident. A young assistant professor named James Olds and graduate student Peter Milner set up an experiment to study learning and attention in rats. They implanted electrodes into the brains of the animals, then put them in a maze. Every time the rats left the maze and approached the edge of the cage, they were given a mild electrical shock. The idea was to get the attention of the rats and encourage them to return to their "studies."

There was only one problem. One of the rats didn't seem to mind the shock. In fact, he kept returning to the edge of the cage again and again, coming back for more.

When Milner and Olds dissected the brain of this peculiar rat, they realized that they had misplaced the electrode. Instead of implanting it in the reticular formation, as they'd planned, they had stuck it into the medial forebrain bundle, where it stimulated dopamine neurons leading into the nucleus accumbens, a group of cells in the basal forebrain. Subsequently, they deliberately implanted electrodes in this region and allowed the rats to press a lever that controlled the current to the brain. Soon the rats were pressing the lever thousands of times per hour—and ignoring levers that released food and water.

As it turned out, Milner and Olds had discovered the brain's pleasure center. Triggering the center causes the release of dopamine in the nucleus accumbens, which is what makes you feel good when you eat a delicious meal, smoke a cigarette, or have sex. Drugs like cocaine, speed, and crystal methamphetamine artificially release dopamine, which is why they are so alluring and so addictive. Drugs that block dopamine have just the opposite effect; they make activities that are normally pleasurable seem unrewarding. The dopamine blockers are all experimental drugs that readers will not be familiar with. For obvious reasons, because they inhibit our reactions to pleasurable stimuli, they don't have much commercial market potential.

The multiple effects of dopamine on behavior have been demonstrated in mice that are genetically engineered to produce either too little or too much of this monoamine in the brain. As you might suspect, mice lacking dopamine are lethargic and uninterested in their surroundings. They become total couch potatoes; they don't interact with other mice, and, by and large, they can't even be bothered to eat or drink. But they recover rapidly when given supplemental dopamine.

The mice with excess dopamine act exactly the opposite. They can't sit still, spending most of the day running around in circles. Life is incredibly stimulating—or at least seems so—even when there is nothing going on.

It is clear from physiological experiments, pharmacological studies, and the powerful appeal of drugs like cocaine that dopamine is just as strong a reinforcer in humans as it is in mice. It seems a reasonable hypothesis that whatever it is that makes spirituality so attractive to people likely involves dopamine.

The last of the three major monoamines is noradrenaline, also known as norepinephrine. Norepinephrine works as the brain's alarm system. When something unexpected or stressful occurs, the amygdala sends an emergency call to the noradrenaline-secreting cells deep in the brain stem. These reply by increasing the output of the monoamine to the forebrain, cortex, and hypothalamus. As a result, the brain is put on code red: Be prepared for trouble. Animals in which the noradrenergic axons are destroyed lose their normal responsiveness to emergencies. They fail, for example, to exhibit the rise in blood pressure that normally accompanies social isolation.

None of the monoamines work in isolation. There is a tremendous amount of "cross-talk" between them. For example, there is evidence that changing the ratio of serotonin to dopamine in the brain results in altered levels of noradrenaline as well, which in turn can act as either a reward or deterrent chemical signal. As a result, drugs that are quite specific at the chemical level can have many different effects on mood, perceptions, and personality. They can also have differing long-term effects on consciousness and spirituality.

Mystical States

The psilocybin that Michael Young ingested on Good Friday 1962 directly affected his brain for only a few hours. It bound to a few serotonin receptors and ever so briefly changed the conformation of some signaling proteins. Then it was broken down and excreted just like waste from food. By the next morning, both his synapses and his perception had returned to normal.

Yet the effects of the experience he had while on psilocybin are still with Young to this day. As he recently commented to a reporter, "What the drug experience did for me involved a deepening of my own spiritual sense, along with a broadening of it. It has influenced the whole context of my ministry."

How could such a temporary change in brain chemistry have such long-lasting consequences? I believe it's because the psilocybin made Young question the one thing that he, like all of us, depends on most in life: his consciousness. Without consciousness, there would be no "life" as we know it. We wouldn't know who we are or where we are going. Yet we take our consciousness for granted. It is automatic and effortless, like breathing. Only when it suddenly stops working and goes "haywire" do we realize what an incredible gift it is.

Episodes of altered consciousness are by no means limited to people taking drugs, of course. Virtually every great religious figure has reported one sort of mystical experience or another. Buddha reached nirvana under the Bodhi tree. Saul was blinded by the light on the road to Damascus. Jesus struggled with the devil in the desert. Muhammad ascended through seven layers of heaven on a winged steed in Jerusalem. Regardless of what one believes actually occurred during their experiences, they all involved altered perceptions of self and surroundings.

Could a gene such as VMAT2 predispose people to such mystical experiences and hence spirituality? To investigate that question scientifically, it is necessary to postulate a mechanistic model of consciousness. In the next few paragraphs, I will present a prototype based on the work of Gerald Edelman that, although undoubtedly preliminary and incomplete, is at least plausible in terms of our current knowledge of neurobiology, and that is scientifically testable. This model assumes that there are distinct processes for core consciousness, higher consciousness, and the integration between them.

Core consciousness, again, is the ability to construct scenes from sensory data. It is core consciousness that psilocybin altered during Michael Young's Good Friday experience. Things around him no longer looked the same; ashes seemed to turn to pearls, and he perceived hair as a halo. Such a flagrant interruption of normal perceptions must have been quite a shock to a young, drug-naive divinity student. It would be, I would think, to anyone.

Higher consciousness is the part of consciousness that integrates the self into our picture of the world. It is the process that maintains our sense of identity. It is higher consciousness that is altered in extreme cases of self-transcendence—creating the feeling of at-one-ness with the world and everyone in it that is so frequently reported by mystics. The ability to "forget" about oneself is the aim of many of the great meditative traditions. This aspect of consciousness, unlike primary consciousness, seems to be unique to humans.

The integration between primary and higher consciousness allows us to remember the present. And this is where VMAT2 enters the picture. VMAT2 controls the flow of monoamines within the brain.

Again, monoamines don't transmit specific information. They can't tell you whether the light is red or green. Rather, they determine the overall tone of the brain: how it feels to be in a moving vehicle, or why it's a bad idea to run a red light. Or, in some cases, what it feels like to encounter God.

People who have had mystical experiences often report a shift in their entire value system. Some of the young divinity students who took psilocybin in 1962, for example, became more involved in humanitarian causes as a result. One of them later commented,

"I got very involved with civil rights after [the psilocybin experience] and spent some time in the South. I remember this unity business. I thought there was some link there." (The overwhelming majority of those who have mystical experiences, of course, do it without the ingestion of artificial drugs.)

For others the change in values is more inward. Tenkai, the young German I met in Japan, had his first *satori*, or enlightenment experience, while weeding the monastery garden one afternoon. "After that," he told me, "everything changed. Everything. I found out that the world was still there even without me. That's when I realized my values were really screwed up. The most important thing isn't whether I have a girlfriend or a nice car or a house in the country. The most important thing is the way that rock looks."

He pointed to a rock under an artificial waterfall in the restaurant where we were lunching. "That rock, the way it glistens. That's a miracle."

Consciousness is an intrinsically private matter. I will never know exactly what happened to Tenkai when he reached enlightenment in the monastery garden—what rendered that one moment in time so very special. But whatever neurological key was turned, whatever electrochemical switch was thrown, it probably involved monoamines.

Seven

How the Brain Sees God

The faith state may hold a very minimum of intellectual content.

—*William James*

One afternoon while we were weeding the vegetable garden at the Hosenji monastery, I asked Tenkai how he meditates. He answered in two words: "I sit."

"Yes, I know," I replied, rubbing my posterior. After sitting for several hours every day on a hard wooden floor, it was difficult to forget. "But how do you keep your mind clear and focused? How do you *not* think?"

"I sit," Tenkai said.

Not satisfied with his simple answer, I asked, "Should I focus on my breathing, or is it better to try to be completely passive?" I asked.

"Try just sitting," Tenkai said. And with that, he flashed his spiritual smile and went back to weeding.

The meditation technique practiced by Tenkai is known as zazen. It was developed 1,600 years ago by Bodhidharma, a devout monk who was the twenty-eighth in a series of masters said to reach back to Siddhartha himself. Bodhidharma traveled from his native India to China, where, according to legend, he spent nine years facing a monastery wall in continuous meditation. His Ch'an school of Buddhism was exported in the eleventh century to Japan, where it became known as Zen.

Zen holds that there is a fundamental unity underlying all experiences and phenomena. It differs from other Buddhist schools and sects in that it teaches that the way to perceive that underlying reality is through pure intuition, not scripture or intellectual learning. The focus is on quiet contemplation rather than ritual, on meditation rather than book learning. "If you wish to seek the Buddha," according to Zen thought, "you ought to see into your own nature, for this nature is the Buddha himself."

The ultimate aim of zazen is to discipline the mind to the point where the practitioner can achieve enlightenment, or satori—a sort of mystical intoxication in which one escapes consciousness of the self and enters into a sense of oneness with all reality. Satori represents a complete reordering of the relationship between self and surroundings. It typically requires a long period of intense preparation, but it may be triggered by a chance occurrence: an accidental splash of hot tea on the hand, the sound of a crow's caw, the sudden fragrance of plum blossoms. Zen annals record that Daigu reached enlightenment on a hot summer day when he chose to perform zazen on a bench suspended over a cooling well. The plank broke, he fell into the well, and he experienced—nirvana.

Although I have never experienced satori, the descriptions from Zen adepts are enticing. Tenkai speaks of a peacefulness beyond imagination. Others find boundless joy. The most common feature is a radical form of self-transcendence. One Zen practitioner reported:

> Ztt! I arrived. I lost the boundary of physical body. I had my skin, of course, but I felt I was standing in the center of the cos-

> mos. I saw people coming toward me, but all were the same man. All were myself. I had never known the world before. I had believed that I was created, but now I must change my opinion: I was never created; I was the cosmos; no individual existed.

It sounds like an altered state of consciousness caused by drugs, yet no external substance was involved. This was a natural high.

How could something as simple as sitting and not thinking lead to such a radical change in a person's consciousness? What could possibly be going on inside the brain?

This Is Your Brain on Zazen

To find out, Andrew Newberg took snapshots of the brain on meditation.

Newberg is the director of nuclear medicine at the Hospital of the University of Pennsylvania. He makes his living by making PET scans of people with brain tumors and neurological disease. But his hobby for many years has been the biological basis of religion and spirituality. Together with psychiatrist Eugene d'Aquili, he has developed a theory about how belief in God is encoded in the neurological architecture of the human brain.

Newberg used a SPECT camera to make his snapshots. SPECT is an acronym for *single photon emission computed tomography*, which is a high-tech method for precisely locating radioactive emission from the brain. By injecting subjects with a radioactive tracer that is carried by the bloodstream, it's possible to determine which parts of the brain are more or less active under particular circumstances or in a particular individual. The advantage of SPECT over other brain-imaging techniques is that the tracer locks into brain cells and remains there for many hours. This means it is possible to make a true "candid" image of the brain—one instant in time frozen for the camera.

The subjects for the experiment were eight experienced Tibetan Buddhist meditators. All of them had been practicing for many years, which gave them the focus needed to reach a high level of absorption even in the unfamiliar setting of a hospital. Fortunately, the design of the experiment allowed them to meditate quietly, alone in a private room with burning incense and candles. Their only connection to the outside world was a piece of string that led to another room where Newberg sat ready with the radioactive tracer. When the meditators reached the peak of their spiritual experience, they tugged on the string and Newberg immediately injected the tracer into a long intravenous line leading into their arms. A few minutes later, they were trundled off to be photographed under the SPECT camera, a large robotlike device carrying a triple-headed detector with whirring magnets.

The resulting snapshots showed what was going on inside the meditators' heads. Areas of high brain activity were colored red, those of low activity came out blue. By comparing before and after pictures of each subject, even the most subtle effects of meditation could be detected.

The most obvious consequence of meditation Andrew Newberg could see was an increased blood flow through the frontal cortex and the thalamus. Many areas that were barely pink in color on the SPECT images at the start of the experiment became red hot after the subjects had meditated for an hour; the dorsolateral prefrontal cortices, inferior and orbital frontal cortices, sensorimotor and dorsomedial cortices—all components of the cortex, the part of the brain responsible for thinking and planning—showed significant increases in activity.

The thalamus and cingulate gyrus, components of the limbic system—the emotional part of the brain—also heated up as the Buddhists meditated.

Newberg and d'Aquili call this aggregate of structures the attention association area, because it is involved in concentration and planning. It is sometimes referred to as the "neurological seat of the will" because of its importance in goal-oriented behaviors and actions.

This increased activity of the attention area was expected. Studies of meditators using electroencephalography, or EEG—an older and less precise method of brain mapping—have also shown increased activity in the frontal region. Remarkably, one study demonstrated that counting silently resulted in increased activity in the attention association area (whereas counting out loud activated the motor areas necessary to move the mouth and tongue). What this part of the experiment proved was that the meditators were doing exactly what they were supposed to be doing. They were focusing.

The second result of the experiment, however, surprised researchers. Several parts of the brain showed *decreased* activity during meditation, as if they were shutting down. The diminution was most notable in the posterior superior parietal lobes, which are located toward the back of the head. There was a strong correlation between increased activity in the left prefrontal cortex and decreased activity in the left superior parietal lobe. The more the Tibetan meditators used the front of their brains, the more activity in the back parts tapered off.

The posterior parietal lobes act as an orientation association area that plays a critical role in defining the self. Information from touch, vision, and hearing is received by this part of the brain and organized into a three-dimensional picture of the body and its position in space. When this area is damaged—by a stroke, for example—the person has problems navigating. These people can't tell where they are—or in some cases even who they are.

Imaging studies suggest that the left and right orientation areas have slightly different functions. The left lobe helps the mind define the limits of the body, whereas the right lobe locates the body within space. The overall effect is to distinguish between your body and all this nonbody—in other words, between self and nonself.

The importance of the orientation association area has been demonstrated through fascinating experiments with monkeys. When monkeys are shown objects within their grasp, one set of parietal lobe neurons fires off, but when they see things outside their reach, a different group of cells is activated. This makes sense, since monkeys logically define "mine" and "not mine" by what they can grab on to.

A slightly different set of brain cells is activated when a monkey actually moves his arms. The same cells fire off when the monkey's own arm is replaced by a realistic-looking prosthesis, but not if it is replaced by an unrealistic substitute such as a mechanical arm. In other words, the monkey's orientation area can distinguish self from nonself even when "self" is actually something else.

A Man Trying to Think Like a Dog

What role does turning on one area of the brain and turning off another play in heightened spirituality and mystical experiences? When I described Newberg and d'Aquili's results to Gerald Edelman, he responded: "That's because a mystic is just a man trying to think like a dog." At first I thought he was joking, but when I considered the SPECT scan data in light of Edelman's theory of consciousness, the comment began to make sense.

Remember, Edelman's theory holds that humans have two layers of mental life: core or primary consciousness, and secondary or higher consciousness. Core consciousness, the formation of scenes from sensory data, is evident in most animals, whereas secondary consciousness, which involves self-awareness, is, as far as we know, uniquely human. Thus, "thinking like a dog" is shorthand for a shift in the balance between primary and secondary consciousness.

The part of the brain responsible for primary consciousness, according to Edelman, is the thalamocortical loop, consisting of the cortex and the thalamus—precisely the same area as the "attention association area" that showed increased activity in the Buddhist monks studied by Newberg and d'Aquili. With these ideas in mind, the SPECT scan results can be interpreted according to the following scenario.

As the Buddhists meditate, they consciously attempt to clear their minds of thoughts and emotions. To do so, they send signals through the thalamus to the cortex, the seat of the will. As more and more of the brain's energy is directed toward this area, output to other regions is decreased through the process that neurobiologists call deafferentation. It would be like simultaneously turning on all the air conditioners in a house; the flow of power to the other appliances would be decreased.

As more and more neural activity is directed to the thalamocortical loop, less and less is available for the orientation association area. The result was a loss of the usual sense of self and space that is conveyed by the posterior parietal lobes. As a result, the Buddhist meditators' brains could no longer tell where their bodies began and the outside world ended. They lost secondary consciousness even though primary consciousness—awareness—was normal or even heightened.

Initially, this deafferentation evidence in the Buddhists was mild—a slight dreaminess. But as the meditation deepened and the power shortage became more acute, the orientation area began to send out SOS calls to the limbic system, thereby activating the cingulate gyrus. The limbic system, especially the hypothalamus, in turn called up the thalamocortical loop, telling it to find out what was going on—which made it work all the harder, resulting in even more deafferentation of the posterior parietal lobe.

Soon the neural phones were buzzing off the hooks with back-and-forth calls about the emergency situation in the orientation association area. Ironically, each call made the situation even more serious as the wires became more and more overloaded. As Newberg and d'Aquili elegantly describe the situation, "A reverberating circuit is established in the brain, with a stream of neural impulses gathering strength and resonance as they race again and again along their neural speedway, fostering deeper and deeper levels of meditative calm with every pass."

This is how the Buddhist meditators were able, in Edelman's more prosaic words, to "think like dogs." By focusing all of their mental energy on primary consciousness, they decreased their secondary consciousness to the point of losing the usual sense of self. They became one with the world and the world became one with them—all through the redirection of a few nanovolts of electrochemical energy.

The effects of spiritual activity on the brain are not limited to Buddhist meditators. Newberg and d'Aquili have also used SPECT imaging to study Franciscan nuns immersed in prayer and observed very similar changes in their scans. The main difference was that the sisters described their peak moment as a "tangible sense of the closeness of God and a mingling with Him," rather than as a "Ztt" experience like the Buddhists. Although SPECT has never been used on Muslims at Mecca or Jews at the Western Wall, it is likely that their brains go through similar changes.

A Spectrum of Selflessness

Newberg and d'Aquili's brain scan experiments offer an intriguing glimpse into how the brain becomes self-transcendent in Buddhist monks and Franciscan nuns, but what about ordinary people? Can an altered balance between core and higher consciousness be achieved by regular folks—and if so, how?

Over the centuries, people have been fascinated with self-transcendent states and have developed many different methods to induce them. Even before written history, cave paintings suggest that shamans were experiencing out-of-body trance states. In the twelfth century, the Sufi mystic Ibn al-'Arabi wrote, "He who knows himself understands that his existence is not his own existence." To Walt Whitman, the ninteenth-century American poet, prayer was "to return from the solitude of individuation into the consciousness of unity with all that is."

It is not just shamans, mystics, and poets who experience a loss of normal self-identity. A Gallup poll reported that more than 40 percent of Americans have had what they consider to be a mystical experience. Another survey, this one conducted by psychiatrists, found that about one-third of Americans have at some point felt "separated from their own self."

Nor is the separation between self and other necessarily limited to mystical reveries. Newberg and d'Aquili argue that there is a gradient of self-transcendence that extends from aesthetic appreciation at one end to Absolute Unitary Being at the other. Each step in this continuum is characterized by increased deafferentation of the orientation association area, resulting in a heightened sense of unity over diversity.

At the low end of the scale are aesthetic experiences such as visiting a cathedral or watching a sunset. Such occurrences often evoke the feeling in one that the whole is more than the sum of the parts. Sometimes the viewer is struck with the feeling that he or she is very small indeed—that the world is much grander than the person had hitherto imagined.

Next up Newberg's scale is romantic love, where you feel totally joined to the object of your affection. It is probably no accident that religious people throughout the ages, from the sixteenth-century St. Teresa to twenty-first-century rapper Lauryn Hill, use words like "rapture" and "ecstasy" to describe their spiritual experiences. It may be significant that sexual drives are controlled by the hypothalamus, the same structure that mediates the emotional value of mystical experiences.

Moving up the continuum is the experience of religious awe, or numinosity—that strange mix of fear and exaltation that Rudolf Otto termed *"mysterium tremendum et fascinosum."* Although such feelings sometimes arise spontaneously, more often they are invoked by sacred rituals and symbols. Newberg and d'Aquili argue that numinosity arises from a mild stimulation of limbic structures—the hypothalamus together with the amygdala and hippocampus—leading to partial deafferentation of the back part of the brain, especially the visual association area.

The most extreme deafferentation occurs in what Newberg and d'Aquili call "Absolute Unitary Being." This is the blissful state of complete union and undifferentiated oneness termed Unio Mystica by Christians and nirvana by Buddhists. It is characterized by a deep sense of harmony and a total obliteration of the self. It is usually achieved by intense meditation, either passive or active, but can also arise spontaneously in some individuals.

Although Absolute Unitary Being is not common—many people never experience it at all—the basic brain process of a shift in the balance between primary and secondary consciousness can occur to some extent in any person at any time. This is part of the universality of spirituality; it's available to everyone. The reason for that universality, Newberg could claim, is that our brain circuits are plastic—the way they handle information can change. A seismic shift can occur as a result of a mystical experience when you are digging in the garden, watching TV, or walking down the street. More often, though, it takes some sort of nudge.

Drumbeats

Often that nudge comes, at least in part, in connection with rhythm: the beat of a drum, the cadence of a chant, the stomp of feet in unison.

Music and dance are among the most ancient and enduring of ritual practices. The reason has to do with more than tradition, for rhythmic activities can promote the same sort of brain activity associated with mystical states.

The Ju/'hoansi, a !Kung-speaking group of hunter-gatherers living in Botswana and Namibia, for example, have incorporated dance and song into their spiritual life in an especially vivid fashion. After sundown, a sacred fire is lit. Female singers arrange themselves in a ring around the blaze and begin a complex, rhythmic melody. Male dancers form an outer ring encircling the women and the fire. They dance for many hours, beating a trough several inches deep into the sandy ground. As dawn begins to break, the dancing and singing intensify. Some of the men enter a trancelike state; their eyes become glassy, their breathing labored, their footwork ever more emphatic. They "leave" their earthly bodies, entering a state of altered consciousness called *!kia*. During this time, it is believed they can heal others of any illness and throw themselves into a hot fire with impunity.

The whirling dervishes, members of the Mevlevi sect of Islam, half chant and half shout the name of Allah as they perform their swirling, leaping, ecstatic dance. The sun dance of the Plains Indians is a major communal religious ceremony, celebrating the spiritual rebirth of the participants as well as of the Earth. Believers in voodoo use dance to invoke spirits from the world beyond. Even when dance is performed in a completely nonreligious, nonspiritual setting such as a nightclub, it has the power to take people outside of themselves.

Electrical recordings of the brain show that "getting the beat" is more than just a phrase—musical rhythms can actually alter brain activity. Andrew Neher, a psychologist at East Los Angeles College, found that brain waves changed in response to loud, rapid drumming. Makoto Iwanaga, a professor of behavioral sciences at Hiroshima University, has studied the effects of different types of music by looking both at electroencephalograms and physiological signs, such as body temperature, skin conductance, and respiration. Excitative music, such as the fourth movement of Tchaikovsky's Fourth Symphony, caused more activation of both the body and the brain than did sedative pieces like the third movement of Mahler's Sixth Symphony. It didn't make any difference which piece the subjects preferred; what was important was the musical content.

Some music, like spirituality, carries an emotional punch. Studies of stroke victims lead to the conclusion that the emotive power of music is independent of its intellectual content. Some people whose speech centers have been damaged by stroke still retain their musical ability. The Russian composer Vissarion Shebalin, for example, completely lost his ability to speak and to understand speech but kept right on composing.

Even more remarkable is the fact that individuals who have lost their musical ability continue to show an emotional response to music. One woman, whose brain was damaged during surgery, became musically incompetent. She could no longer sing or even recognize a simple tune like "Mary Had a Little Lamb." Nevertheless, she reported that music still made her feel happy.

To test this seemingly implausible claim, doctors hooked her up to devices capable of measuring various physiological parameters, then played different types of musical selections: major and minor chords, fast and slow pieces, and so on. Sure enough, her emotional responses were completely normal. The exciting pieces excited her, and the calming pieces calmed her—all without her being able to recognize a single tune.

Positron emission tomography, known as PET scanning, has shown that the part of the brain involved in this type of musical appreciation is the temporal lobe. Remarkably, this same region plays a key role in another trait in which there is often a dissociation between intellectual understanding and emotional impact: spirituality.

Electrical Storms

What did the apostle Paul, Muhammad the prophet, Joan of Arc, and Fyodor Mikhailovich Dostoyevsky have in common? All of them were intensely religious. All of them had mystical visions. And some scientists now think all of them may have owed at least part of their intense feelings of spirituality to temporal lobe epilepsy, a neurological disease that causes abnormal electrical firing in the limbic system.

Temporal lobe epilepsy is a common disorder, affecting about 2.5 million Americans. It is characterized by dreamy, hallucinatory seizures. Suddenly, for no apparent reason, the person becomes motionless and unresponsive. They seem frozen, but inside their heads the world is spinning. They may have visual alterations; lights seem to flash, and objects appear clearer or blurrier than normal, nearer or farther away. These visual effects are often accompanied by auditory illusions, including voices. Other sounds seem louder or softer than usual, or nearer or farther away. There also are emotional alterations; strong feelings of sorrow, joy, fear, or disgust well up without reason.

The most common sensation in temporal lobe seizures is that things just aren't quite the way they normally are. One psychiatrist had a patient who spent long periods staring at her coffee table because "it just doesn't look exactly like my coffee table." She didn't think the table actually changed, just her perception of it. Her consciousness was altered.

Physicians often miss temporal lobe epilepsy because it does not cause the dramatic convulsions of a grand mal seizure. The only way to make a sure diagnosis is for the person to spend weeks hooked up to an EEG machine in a hospital ward. Obviously, we don't have electroencephalographs for Paul, Muhammad, Joan of Arc, or Dostoyevsky. There is, nevertheless, anecdotal evidence to suggest that each of them had temporal lobe epilepsy.

Paul had his most famous seizurelike experience when he was still Saul, a Jewish Pharisee traveling on the road to Damascus. According to the Bible, he saw a flash of light, fell down, and heard the voice of Christ. Afterward he was blind for several days. This episode has all the earmarks of an epileptic episode: Loss of balance, visual illusions, and auditory hallucinations are common in temporal lobe seizures. Blindness is rarer but has been reported in some cases.

If Paul had experienced just one episode on the road to Damascus, the diagnosis of temporal lobe epilepsy would be sketchy. But apparently it was part of a pattern. Paul later reported on an otherworldly trance that included "visions and revelations." His fellow apostle Luke reported that Paul had a "bodily weakness." In fact, many biblical scholars, including William James, believe that Paul suffered from epilepsy. His conversion experience was part of a persistent pattern, not a fluke.

Muhammad also had many seizurelike experiences; he saw flashing lights, heard the voices of the angel Gabriel and Allah, and suffered from fits of trembling and profuse sweating and bodily pain. He also had several out-of-body experiences, a common feature of temporal lobe epilepsy. Again, it was part of a lifelong pattern. Legend has it that Muhammad was born with excess fluid around the brain and had fits as a child.

Flashing lights and mysterious voices are two of the most common hallucinations in temporal lobe epilepsy. Joan of Arc experienced both:

> I heard this Voice to my right, towards the Church; rarely do I hear it without its being accompanied also by a light. This light comes from the same side as the Voice.

Joan had many such conversations with God, which guided her battle strategies.

Although Dostoyevsky is best known as a novelist, he was also an intensely spiritual person who believed that "man seeks not so much God as the miraculous." His interest in mystical experiences may have resulted from his epilepsy, which is documented in statements by his physicians, several biographies, and his own voluminous writings. Many of his fictional characters were tormented by their seizures, but Dostoyevsky enjoyed his—or at least the beginnings of them:

> You strong people have no idea of the bliss which epileptics experience in the moments preceding their attacks. For several moments, I have a feeling of happiness which I never experienced in my normal state and which one cannot imagine. It is a complete harmony in myself and in the wide world . . .

The effects of temporal lobe epilepsy are not limited to the brief storms of electrical activity that occur during seizures. There are more abiding consequences. As noted by necrologist and psychiatrist Norman Geschwind, temporal lobe epileptics display many characteristic personality features even in the intervening periods between seizures. They often express an intense interest in philosophical issues and write about them extensively, a trait called hypergraphia. Sometimes they change their sexual preferences or lose their sex drive altogether. There is a tendency to imbue even ordinary scenes and events with emotional significance.

Most of all, they are hyperreligious. They attend religious services twice a day, build shrines in their homes, and have long conversations with God. They may give up their jobs and ignore their families to pursue their religious interests. They become zealots.

Now, clearly not every religious person has temporal lobe epilepsy—far from it—and not every temporal lobe epileptic becomes obsessed with religion. But the connection between the two is strong enough to make scientists wonder how it arises in the human brain.

The God Spot

Michael Persinger, a professor of psychology at Laurentian University in Canada who specializes in paranormal phenomena, thinks he knows the answer. Persinger had always been a nonbeliever. Then, in the course of using transcranial magnetic stimulation to study the function of various brain regions, he stimulated his own temporal and parietal lobes. For the first time in his life he experienced God. He had hit "The God Spot."

Based on this experiment and other lines of evidence, Persinger believes that the biological basis of all spiritual and mystical experiences is due to spontaneous firing of the temporoparietal region—highly focal microseizures without any obvious motor effects. He calls such episodes transients and theorizes that they occur in everybody to some extent. Exactly how often and how strongly is determined by a mix of genes, environment, and experience. The main effect of such transients is to increase communication between the right and left temporoparietal areas, leading to a brief confusion between the sense of self and the sense of others. The outcome, he says, is a "sense of a presence" that people interpret as a God, spirit, or other mystical being.

To test his theory, Persinger outfitted normal volunteers with a special helmet equipped with four sets of solenoids. He seated them in a quiet room, then stimulated their temporoparietal areas with a magnetic field using either a biologically relevant waveform that mimicked the brain's own magnetic activity or, as a control, an irrelevant waveform that would not be expected to have any biological effect. The subjects were asked to press a button if they felt "a presence."

The results were just what Persinger predicted. The volunteers were significantly more likely to press the button when they received a biologically meaningful magnetic field than when they received no radiation or an irrelevant waveform. Some of the subjects even had mystical experiences. One female student reported a "feeling of rising" in which she was floating away into space, retained only by a thin thread.

Although Persinger's stimulation experiments are fascinating, they do have several flaws. One is suggestibility. Perhaps the subjects felt "a presence" only because they were told they might. This is especially troublesome since the volunteers were first-year psychology students who received a grade bonus for participating. ("Yeah, I felt the presence of God. Now do I get an A?")

The second problem is specificity. Does the experiment show that there is actually a "spot" or circuit in the brain that is devoted to spirituality, or does it simply reflect the side effects of stimulating regions normally involved in emotionality and self-recognition? It is hard to tell.

Neuroscientist V. S. Ramachandran took a different tack to find the "God spot." He hooked up electrodes to the hands of two temporal lobe epileptics with religious obsessions, showed them images and words on a computer screen, and recorded their skin response. When the subjects were shown familiar objects, like a picture of their parents or the word "shoe," there was no reaction. The same was true for portraits of strangers, erotic pinup shots, and four-letter words. Even the sight of a man being eaten alive by an alligator failed to get a response.

But when the temporal lobe epileptics were exposed to religious words and images, their skin responses went through the roof. Just seeing "God" on a computer screen was enough to set them off. All it took were a few well-chosen words and pictures to turn on the God spot in these individuals. By contrast, nonepileptic individuals showed just an average response to the religious symbols, reserving their strong responses for the sexual and violent scenes.

Ramachandran concluded that epilepsy caused permanent changes in the temporal lobe circuitry. Some circuits were enhanced, others diminished, leading to "new peaks and valleys in the patients' emotional landscape." The importance of his experiment is that it shows that the temporal lobe response is specific for religion. The electrical firestorms did not simply cause a general enhancement of emotion (a phenomenon called "kindling"); if that were the case, the epileptics would have responded more *strongly* to sexual and violent scenes, rather than more weakly. There was something about the way that God—or even the idea of God—made them feel that changed in their brain.

The Way God Feels

Over the past three chapters, I have covered several topics: genes and phenes, monoamines and vesicular transporters, primary and secondary consciousness, the temporal and parietal lobes and the connections between them. Yet the natural question is How does this mix of biochemistry, neurology, and anatomy help to explain the existence of a God gene and its role in spirituality?

Let me recap the three central arguments of the God gene theory. First, the sense of self is central to spirituality. Whether it is the lifelong trait of self-transcendence or an ephemeral experience like Daigu's nirvana in the well, the result of years of meditation or the effect of popping a pill, the hushed reverence of a church choir or the frenzied exuberance of African drums, traditional or New Age, ritualistic or free-form—in every case the ability to lose one's sense of self, to become at one with the universe and everybody and everything in it, is at the heart and core of spirituality.

Second, our sense of self and of the world around us arises from the distinctive brain process of consciousness. Although we do not yet understand all the details, it appears that higher consciousness, or "me-ness," depends on structures in the back of the brain that are involved in orientation, whereas core consciousness, the ability to construct scenes from sensory data, is encoded by the more frontal structures of the thalamocortical loop and attention association area. These circuits are interconnected yet distinct.

Third, monoamines play a central role in consciousness by lending value to perceptions. It is the monoamines that make us feel good, bad, or somewhere in between about other people, places, and experiences. Such evaluations are essential to our mental life; without them, there would be no meaning to what we do or experience. By influencing the ebb and flow of monoamines, VMAT2 helps to determine how we perceive alterations in consciousness.

Consider two individuals sitting side by side on the polished wooden floor of a Zen monastery in Japan or on a pew in a church in New England. As the meditation bell rings or the doxology is sung, both people focus all of their mental energy on the association area of the brain. This results in a partial shutdown of the orientation area that is mediated by the thalamus, which in turn sends off signals to the limbic system via the hippocampus, amygdala, and hypothalamus, resulting in varying degrees of mental excitement depending on the individual.

In one individual, with one particular version of the VMAT2 gene, these signals result in only a modest alteration of monoamine signaling. The call is received, but it's not urgent. It doesn't feel important. Perhaps the person's mind drifts to the eternal question of why monastery floors and church pews are always made of such hard wood, or what they are likely to be having for lunch.

In another individual, with a different VMAT2 gene, the same signals have a more dramatic effect because they are received by a different monoamine transporter. Serotonin, dopamine, and noradrenaline rush in and out of this individual's vesicles, setting off a reverberating circuit that is accelerated by ever stronger signals from the cortex and ever weaker input to the parietal lobes. The result is a radical shift in the communication between the front and back of the brain—a shift that, in this individual, brings a profound sense of joy, fulfillment, and peace.

This is why feelings of spirituality are a matter of emotions rather than intellect. No book or sermon can teach one person to use a different monoamine transporter or another to ignore the signals emanating from his limbic system. It is our genetic makeup that helps to determine how spiritual we are. We do not know God; we feel him.

Eight

Evolving Faith

If we were to ask the question: "What is human life's chief concern?" one of the answers we would receive would be: "It is happiness."

—*William James*

Where did God genes come from?

At first, that might seem like a question of faith or philosophy rather than science. But the actual answer is quite obvious. They came from our parents, who inherited them from their ancestors. Those ancestors received them from their predecessors, and so on down the evolutionary line to the very beginnings of life on Earth.

Over the ages, of course, the genes evolved. At every step, the genes that helped their owners survive and reproduce were most likely to be passed on to the next generation. Genes that don't successfully accomplish this don't survive in succeeding generations. If the organism in which such genes resided didn't have offspring, the genes soon would be lost from the population, discarded in the dustbin of failed evolutionary experiments. Only the genes that promoted our past survival and reproduction are still with us today.

How did God genes evolve? What advantage do they confer to humans?

We do not yet have definitive answers to those questions, and it is possible we never will. Sociobiology and evolutionary psychology—the two disciplines that analyze the evolution of human behavior—are inexact sciences because we have no way to re-create evolution in the laboratory. Nevertheless, we can make educated guesses based on what we know, including the new research revealed here about God genes and monoamines.

The Challenge

Evolutionary biologist Edward O. Wilson considers religion to be "the greatest challenge to human sociobiology and its most exciting opportunity to progress as a truly original theoretical discipline." While I agree, so far science is still struggling with the challenge.

Wilson invented the field of sociobiology. Kindly, soft-spoken, and unfailingly courteous, he sometimes seems out of place in the elitist environment of Harvard, where he has studied ants and taught evolutionary biology for more than four decades.

In his groundbreaking book *On Human Nature*, Wilson lays out the evidence that the predisposition to religious belief has a genetic basis. There is evidence of religious belief more than 60,000 years ago among Neanderthal man. In fact, it is universal; every society, from hunter-gatherers to postindustrial democracies, has had some form of spiritual belief.

Wilson proposes multiple mechanisms for the evolution of God genes. Genes that make people open or susceptible to religious indoctrination may have evolved through selection of clans competing against one another. Other genes might alleviate anxiety by allowing people to become members of a cohesive group. Wilson even suggests that some religious traditions have direct selective advantage; the kosher laws of Judaism, for example, he argues, might have protected people from food poisoning in the days before refrigeration.

Richard Dawkins is another pioneering evolutionary biologist, best known for popularizing the concept of "the selfish gene," but he doesn't think that genes have anything to do with spiritual beliefs.

Dawkins, who speaks in the clipped, aristocratic accent of an Oxford don, is aloof, a tad pompous, and infamously intolerant of fuzzy thinking. He considers religion to be "a virus of the mind"—a parasitic group of myths and falsehoods that serve no biological function or advantage. These culturally transmitted ideas, or memes as he calls them, have no purpose other than to copy themselves. (I will return to the role of memes in religion in the next chapter.) Dawkins argues that genes play only an indirect role in religion and spirituality by facilitating the formation of brains that can transmit cultural information.

How can two such distinguished scientists hold such diametrically opposed views? Part of the difference in their views may be a result of their backgrounds. Wilson, who was raised in a religious family, appreciates the human value of spirituality. Dawkins, by contrast, is an unapologetic atheist who seems to dislike religion intensely.

However, the deeper problem for the two as scientists is that both Wilson and Dawkins have set their sights so broadly. Both of them attempt to derive single explanations for multiple phenomena, mixing together purely cultural traits (like not eating shellfish and pork in the case of Orthodox Jews) with those that may have a more biological basis (such as a willingness to submit to a dominant leader). The two traits are as dissimilar as apples and oranges. If we want to understand the evolution of genes that influence our need to believe, then we need to focus on those traits known to involve genes.

Faith Healing

Magic is the best theology, for in it true faith is grounded.
—*Jacob Boehme*

I believe our genetic predisposition for faith is no accident. It provides us with a sense of purpose beyond ourselves and keeps us from being incapacitated by our dread of mortality. Our faith gives us the optimism to press on regardless of the hardships we face. But does faith meet more than just psychological needs? Might it affect physical aspects of life as well?

One popular concept is that religion helps societies organize and successfully compete against others. But if such group-level selection were the only selective force, God genes would probably die out or be limited to only certain parts of the world, since the necessary conditions—high degree of kinship within the group and high degree of competition with outside groups—are limited to particular geographical areas and certain historical times. To be a universal facet of our evolution, there must be additional reasons to account for the persistence of God genes.

Perhaps one needs to look no farther than religious texts to find an answer: the healing power of faith. One selective advantage of God genes may arise from the ability of faith to improve human health and prolong life.

Consider first the historical evidence. Virtually every great religious founder was also a healer. Jesus of Nazareth is a prime example. According to the Bible, Jesus was an infectious-disease expert who could cure advanced leprosy. He was also a skilled ophthalmologist, capable of treating blindness with a remedy made from his own saliva. With his help, a paralyzed man jumped to his feet and walked for the first time. Others were cured of hemorrhaging and psychosis. In the case of Lazarus, Jesus pulled off the ultimate in medicine: He brought the dead back to life (as he would later be resurrected himself).

Other religious traditions include similar stories of healing and miracles. The Hindu divinity Krishna cured leprosy, helped a lame woman to walk, and brought the dead back to life. Buddha healed the deaf, blind, and sick, including a man who had both feet cut off by an unjust king. Thousands of years ago, Horus, the Egyptian savior, raised the dead to life. Asclepius, an ancient Greek healer who became a demigod, became so skilled in surgery and the use of medicinal plants that he could make malignant tumors regress. His symbol, a snake coiled about a staff, today is the symbol of medicine.

Even in present-day hunter-gatherer societies, religious specialists have diagnostic and healing powers. The Navajo hand-trembler places corn pollen on his patient and himself, prays and makes offerings to the gods, then sings a sacred song. His arm and hand begin to tremble uncontrollably. The duration and direction of the tremors allow him to pinpoint the source of the patient's problems. The sorcerers of Dobu, an island in the western Pacific, divine the source of medical problems by gazing into a bowl of water. The Lovedu people of South Africa throw bones for the same purpose, while the Nyoro of Uganda interpret scattered cowrie shells.

The healers of preliterate societies practice healing rites that are similar to those of modern medicine, including the use of naturally occurring herbal pharmaceuticals and certain forms of surgery. More often, however, their treatments are based on supernatural beliefs and faith. The curers of the Sia Indian tribe of New Mexico believe that witches cause disease by stealing a person's heart. In response, they don a necklace of bears' paws and go into the desert to recover the patient's heart, which they bring back in the form of a ball of rags wrapped around a kernel of corn.

Kapauku Papuans of western New Guinea consider disease to be caused by evil spirits, which they entice to leave the body by roasting small birds and rats over a fire. In the Pacific Northwest, Achomawi shamans go into a trance state, consult with spirits, then lie down next to the person who is sick and suck the affected area. The cure is completed by removing a feather from the affected area, then making it disappear by blowing on it.

Belief in faith healers is by no means restricted to so-called primitive peoples. Christian Scientists are a religious denomination founded in 1879 by Mary Baker Eddy, a New England woman who suffered chronic spinal problems. At the age of 41, she was successfully treated by Phineas P. Quimby, a mesmerist (hypnotist) and spiritual believer who held that happiness is determined by belief. Unfortunately, she suffered a relapse after falling on an icy sidewalk. During her recovery, she came to realize that health comes not from the body or even the mind but directly from God, "the great I AM; the all-knowing, all-seeing, all-acting, all-wise, all-loving, and eternal; Principle; Mind; Soul; Spirit; Life; Truth; Love; all substance; intelligence."

Christian Scientists depend on prayer rather than modern medicine to treat diseases, from the common cold and bellyaches to HIV infection and AIDS. (Broken bones and toothaches, however, merit conventional doctors.) They often recruit the assistance of a Christian Science practitioner—a person who devotes all their time to helping others through prayer. The practitioner does not claim personal healing power but rather acts as an intermediary, asking God for guidance and help. Patients do not need to know the healer personally; several practitioners advertise and practice through the Internet.

The list of kinds of faith healing is extensive. Similar stories are told of the Teleut shamans of the northern steppes of Russia, the Zuni medicine men of the pueblos of New Mexico, the Taoist priests of China, and many more. Does faith healing actually work? We simply do not know. There is no systematically collected data available. The point is that enough people believe that it works to warrant scientific investigation. There is smoke. Is there fire?

Religion and Health

One simple way to investigate this is to explore whether or not religious people live longer. Several studies suggest they do.

For example, an analysis of 3,968 older adults living in the Piedmont of North Carolina between 1986 and 1992 found that 37 percent of those who attended services less than once a week had passed away during those years, while only 23 percent of those who went to church at least once a week died. The difference held up after taking into account age, race, social contacts, and health practices. The frequent attenders had a relative hazard of dying that was 46 percent lower than that for less frequent attenders. The effect was strongest for women but was significant in men as well. Frequent churchgoers lived longer.

Many other studies have yielded similar results. A canvass of 21,000 Americans found that those who never went to church had a 50 percent increased rate of mortality compared to those who attended frequently. Another large survey, which followed 5,286 Californians for 28 years, also found that regular church service attenders had lower death rates. A five-year study of 1,931 older residents living in Marin County, California, gave similar results. Other studies have shown that religiously affiliated individuals have lower rates of heart disease, hypertension, and cancer—the three leading causes of death in the United States. They are more likely to survive open-heart surgery, as well.

When scientists conducted a meta-analysis of 42 independent samples, which included nearly 126,000 participants, religious involvement was consistently associated with lower mortality. The strength of the correlation varied, but the trend was always in the same direction. From the numbers, it appears that a 20-year-old who goes to church regularly stands to live about seven years longer than a similar individual who does not attend services. There seems to be a genuine connection between religion and health.

The question scientists must answer is why. What are the source and direction of the correlation? One possibility, of course, is that religious organizations attract people who have good health habits (e.g., those who do not smoke cigarettes) in the first place. In that case, the association would really be a matter of healthy people being religious, rather than religious people being healthier. The other possibility is that religious involvement actually promotes such behavior, or at least helps people to maintain healthier habits.

To find out the answer, scientists looked at more than 28 years of accumulated data from an Alameda County study, a longitudinal analysis of nearly 7,000 people between the ages of 17 and 95 when the study started in 1965. Over the next three decades, the participants were periodically surveyed about all types of behaviors, from cigarette smoking and alcohol consumption to how many friends they had and how often they attended church.

The results showed that not only did the religious subjects have healthier lifestyles and live longer, they continued to improve over the three decades of the study. Those who went to religious services regularly were less likely to smoke than the nonattenders, and they were more likely to quit if they were already smokers. They also were more apt to work out, get married, and stay married and to refrain from alcohol consumption. They were even more likely to develop a greater number of friendships. The take-home lesson was clear: Being religious is not just a correlate of good health—it actually promotes it.

Such findings suggest that one possible evolutionary advantage of genes that promote our need to believe in something greater than ourselves—what I call "God genes"—is to improve health and longevity. If people with genes that promote faith and spirituality are less likely to become ill and die, they are more likely to pass on these genes to their offspring.

However, there are two limitations to this conclusion. First, frequency of religious attendance—a cultural phenomenon—is only a distant surrogate of what we are really interested in studying, faith and spirituality; in fact, twin studies show that genes play only a minor role in how often people go to church. Second, all of the studies described above looked at people at the end of their life spans rather than during their reproductive years. From the point of view of evolution, it doesn't matter how long you live—just how many children you have. The health studies are a start in that direction, certainly, but we need to look further.

The Power of Prayer

In July 1872, the *London Contemporary Review* published an article titled "The Prayer for the Sick: Hints Towards a Serious Attempt to Estimate Its Value." The author, who declined to be identified, proposed a scientific test of the healing effects of prayer. Two hospitals treating a disease with a known mortality rate were to be identified. Faithful people would pray for the recovery of patients in one of the hospitals, but not patients in the other. After three to five years, the power of prayer would be judged by comparing death rates in the experimental and control hospitals.

The proposal was not well received. The *Review* was flooded with angry letters arguing that it was unethical to pray for one group but not another, that the author didn't understand the purpose of prayer, and that is was silly to rely on observation or statistics when intuition so clearly indicated the efficacy of belief. (Francis Galton, the inventor of the twin method of genetic research, was one of the few advocates of the proposal; he had already shown that members of the royalty did not have an extended life span despite the plethora of prayers offered in their behalf.)

And so the experiment recommended in the *London Contemporary Review* was never carried out. Had it been a modern-day grant proposal, it would not have been funded. Scientists were wary of performing an experiment for which they would have difficulty explaining even a statistically significant positive finding by accepted theories of natural law. And religious believers did not need experimental evidence; they already accepted the efficacy of prayer.

It wasn't until more than a hundred years later that science began to systematically examine the power of prayer. In one of the first published studies of its type, cardiologist R. C. Byrd examined the effects of prayer on 393 patients who were admitted to the coronary care unit of the San Francisco General Hospital over a 10-month period. Byrd assigned 192 patients to an intercessory Christian prayer group outside the hospital, and the other 201 patients to a control group with no supplemental prayer group (although, of course, they may have had family, friends, and their church congregations praying for them). The patients knew they were participating in the study, since they had to sign an informed consent form, but they did not know to which group they had been assigned.

The results were, in a word, miraculous. The patients in the intercessory prayer group did significantly better than those in the control group. They didn't need to be put on a ventilator as frequently, and were less likely to require antibiotics or diuretics. They even got out of the hospital sooner.

It was another 10 years before the experiment was repeated, this time in a hospital in Kansas City. This study included 970 patients, 446 of whom were randomly assigned to the prayer group and 524 as controls. The outside intercessors prayed for a speedy recovery every day for four weeks for those patients in the prayer group.

This time, the results were more ambiguous. Patients in the prayer group stayed in the coronary unit of the hospital just as long as those in the control group. They did, however, have a somewhat less severe course of disease. A review of their charts by "blinded" physicians, who didn't know which patients were in the prayer group and which were in the control group, gave overall severity scores some 10 percent lower than those for the control patients. The authors concluded that "prayer may be an effective adjunct to standard medical care."

The emphasis here is on the word "may." Although both studies seem to show some positive effect of prayer, they are quantitatively minor. The point is that even if intercessory prayer does have some effect on health, this phenomenon by itself is unlikely to explain the existence of "God genes." This is especially true given that any benefits are conferred upon the recipient of the prayers, rather than on the intercessor. We need to look elsewhere for an explanation of the existence of "God genes."

A Miracle Cure

Hurry, hurry, use the new drug before it stops healing.

—*Anonymous*

Why do the miracles reported throughout religious history seem so rare today? For one thing, it may be because people rely more on doctors and drugs than on religious leaders and faith.

To explain what I mean, let's look for a moment at the placebo effect. Placebos are nonactive sham treatments, such as sugar pills or water injections, that are given as controls in clinical trials of new drugs or treatments. "Placebo" literally means "I shall please" in Latin; it was the first word of the medieval hymn sung for the dead. It came to mean "false consolation" because the hymn was often sung by paid mourners rather than family or friends.

The surprise is that placebos often work. The placebo treatment can alleviate symptoms or even cure a disease—as long as the patients believe they might be getting the real drug or treatment.

The most obvious and consistent placebo effects have been recorded for psychiatric drugs such as antidepressants. For example, in the original clinical trials of Prozac, the placebo effect was responsible for fully 50 percent of improvement, as compared to just 25 percent of improvement caused by the drug; the other 25 percent was due to spontaneous remission.

Such strong placebo effects are the rule rather than the exception for antidepressants. A review of all antidepressant clinical studies submitted to the FDA—a requirement for drug approval in the United States—showed that 30 to 50 percent of people get better just by taking sugar pills. In other words, placebos are about 60 percent as effective as the actual drugs. (These substantial effects may actually be underestimated, since many of the studies used inert placebos rather than active placebos that mimicked the side effects of the drug, potentially permitting patients or physicians to figure out who was getting or not getting the actual drug, thus breaking the blind and causing bias that moderated the placebo effect.)

Remarkably, the most serious outcome of depression—suicide—was actually a little lower in people taking placebo than in people receiving the real drug.

Placebos also are strikingly effective in pain management. In fact, they are almost as effective as most over-the-counter analgesics, like aspirin, and 55 to 60 percent as effective as prescription medications such as Darvon and codeine.

The effect of placebos is not limited to pain management and psychological health. Placebos have also been effective for ulcers, rheumatism, dysmenorrhea, herpes, and asthma—not to mention seasickness, acne, migraines, and various neurological conditions. Placebos, like the prescription drugs they are paired with in studies, can even replicate a drug's side effects, such as nausea and fatigue.

Measuring placebo responses can be tricky, since they are not mediated by a "drug" in the usual sense. Rather, they probably represent a mixture of expectation and the therapeutic effects of contact with a caring human being. For this reason, it is crucial to contrast the placebo treatment with the natural course of the disease in the absence of any intervention or medical contact. Simply comparing patients who went to the doctor and received a sugar pill to patients who went to the doctor and didn't get any pill is not sufficient, because it discounts the crucial role of the physician.

The ubiquity of placebo effects suggests that an individual's belief that it can work is enough to make it work on many different types of illness. One has to wonder: If faith in a sugar pill or the medical intervention of a doctor can reduce sickness and prolong life, why can't faith in God, Muhammad, a priest, or a rabbi? And if that is the case, couldn't there be a genetic component that would increase such faith, making it all the more likely such genes would survive and be passed on? But first we need to understand how placebos work. What is the mechanism? Do placebos work solely through the brain, or is there something else going on?

The Monoamine Connection

The first signs of Parkinson's disease are subtle. An author notices that certain keys on his computer keyboard are hard to press. A seamstress has difficulty threading her needle.

Slowly but surely, the symptoms worsen. The slight weakness in one finger spreads to the entire hand, then to the arm and a leg. The minor tremble becomes a constant palsy that can explode into spastic fits. One's gait becomes hesitant, then slows to a shuffle. Speech becomes slurred, and the face becomes frozen and expressionless. Toward the end, even simple daily functions like getting dressed and eating become impossible.

These symptoms are due to the loss of a single brain chemical: dopamine—a monoamine involved both in motor functions and in reward pathways. As the disease progresses, the cells producing dopamine in the striatum and substantia nigra gradually die off. By the time the symptoms are severe enough to warrant a visit to a doctor, only 10 to 20 percent of the normal number are left. The underlying causes of the disease are not precisely known, but may include faulty genes, internal or external toxins, and viral infection.

The first treatment for Parkinson's disease was a drug called levodopa, a precursor of dopamine. In some patients it brought about a miraculous cure—but it also led to debilitating side effects, including nausea, dizziness, and confusion. An improved version that is less rapidly metabolized and thus requires lower doses is now available, but it is still a toxic medicine that can cause as many problems as it solves.

However, there is one treatment that is almost as powerful as levodopa, without any side effects at all: sugar pills. Carefully controlled clinical trials have consistently shown that the tremor, slowness, rigidity, and balance problems of Parkinson's disease all respond to placebo.

How could sugar pills affect something as obviously physical as Parkinson's disease? It has been assumed that the placebo response is indirect rather than chemical. But a group of Canadian scientists decided to check to see if dopamine was involved. They used PET scans to measure the levels of dopamine in the brain. The patients they studied were controls in clinical trials of Parkinson's disease drugs.

The placebo worked. Every patient who was given sugar pills started churning out more dopamine throughout the striatal region of the brain. Remarkably, the improvement was comparable to that seen with levodopa or other prescribed drugs. Moreover, there was more dopamine released in those patients who took the placebo than in those who did not.

Now, this experiment doesn't mean that the "real" drugs don't work for Parkinson's disease. Other experiments in both humans and animals have shown that the well-known beneficial effects of levodopa and similar pharmaceuticals are indeed due to their ability to alter dopamine levels, and that this occurs by a mechanism different from the mechanism by which the sugar pills work. So if you or a friend has Parkinson's disease, don't discount the drugs. Placebos work *only* if you don't know they are a placebo.

The release of dopamine almost certainly is not the only result of the placebo effect. Work by Herbert Benson's group at Harvard, Arthur and Elaine Shapiro at Mount Sinai, and many other scientists has implicated numerous pathways that are affected, including endogenous opiates and nitric oxide. There probably are as many different ways for sugar pills to make people better as there are for prescription drugs. The point is that placebo effects are not "just psychological." They are physical—real—as well.

Nocebos—the Placebo's Evil Twin

As faith can cure, so can it harm. The nocebo is the opposite of the placebo. The nocebo effect is the causation of sickness, or even death, purely by expectation.

Anthropologists have reported an unusual example of the nocebo effect among the inhabitants of Dobu Island off the southern shore of eastern New Guinea. If a Dobu man believes that another person has caused one of his relatives to become sick or die, he chews ginger to make his body hot, drinks seawater to parch his throat so he won't swallow his own evil charms with his saliva, then hides in a tree and waits for his victim. The man jumps out of the tree with a bloodcurdling shriek, brandishing a magic lime spatula with which he makes as if to remove the person's organs. The victim becomes raving mad, never eats again, and soon loses strength and dies.

This Dobu tale is one of many examples of "voodoo death" that anthropologists have collected from various parts of the world, including Haiti, Africa, Australia, New Zealand, and Polynesia. While it is difficult to verify such claims, they offer circumstantial support to the fact that faith—belief—can have a dramatic effect on one's self, and on others who share that belief.

Modern medicine has produced many less-vivid but better-documented examples of the nocebo effect. In one study of individuals with nonspecific chest pain, half of the participants were given the results of cardiological testing that demonstrated they did not have any heart problem, while the other half were not given such test results. The subjects who were kept in the dark fared worse than those who were told that they were fine; they were 2.3 times more likely to report a short-term disability. This means that simply suspecting they had heart disease made the patients feel as if they had heart disease.

Dr. Kenneth Pargament and colleagues at Bowling Green University in Ohio have reported a direct tie between religious anxiety and mortality. They followed 596 elderly hospitalized patients for five years. Among their findings were that those who wondered whether God had "abandoned them" or who "questioned God's love" were more likely to have died over the five-year period than patients who did not report such worries. "We know from quite a bit of research that religion can be a potent resource," Pargament said. "But it's also clear that religion has a darker side. It can be a source of solutions, but it can also be a source of problems."

The effect of negative expectations on health is most clearly documented by the well-known association between mortality and depression. In one of many such studies, scientists at the Centers for Disease Control analyzed the relationship between ischemic heart disease and sadness and helplessness in 2,832 subjects who completed a well-being questionnaire. In order to avoid the possibility that subjects were depressed because of illness, rather than ill because they were depressed, anyone with heart disease at the start of the 12-year study was excluded.

The results showed that people who felt depressed and helpless were 1.6 times more likely to develop ischemic heart disease than those who felt cheery and confident; they were also 1.5 times more likely to die of it. Extrapolated out across the population, that translates to more than 25,000 deaths a year in the United States, or about 1 percent of all mortality. All by feeling bad.

The brain mechanism of the nocebo effect has not been extensively studied. Unlike placebos, which are routinely used in clinical trials, there are few chances to study nocebos. Most medical-subject review panels would rightly look askance at experiments in which the subjects were falsely told they were ill or about to die. Nevertheless, from what we know about depression, there is one logical neurochemical transmitter that might be involved: serotonin.

Remember, serotonin is the "feel-bad" chemical of the brain. Shifts in serotonin levels have been implicated in many different aspects of negative emotions, including fear, anxiety, hostility, and pessimism—all of the personality traits that epidemiological studies find to be associated with poorer recovery from disease and shortened life span.

It has also been shown that serotonin levels can be altered by expectation. For example, in vervet monkeys there is a clear relationship between serotonin levels measured in body fluids and social dominance. The pack leaders have high levels of serotonin, whereas their followers have low levels. When the troop is artificially reorganized so that the top-ranked males are suddenly at the bottom, their measurable serotonin levels plummet, and they become hostile and irritable. Meanwhile, the new leaders gain serotonin and become calm and confident. The changes in serotonin levels did not cause the change in social status. They happened because of it.

The same phenomenon occurs in humans. Fraternity house leaders have different serotonin levels than do new pledges—a situation that is reversed as the ex-pledges become pledge masters, while the ex-pledge masters struggle to enter the workforce. It is one more example of the many two-way streets between chemicals and experience in the human brain.

Taken together, these observations suggest that one way nocebos and negative expectation work is through the increased or decreased release of serotonin. Given the wide range of this monoamine's effects on both the brain and the body, it is perhaps not surprising that belief alone can have such a powerful influence.

The Selfish Spiritual Gene

But to return to the question that opened this chapter: What is the selective advantage that the "God genes" confer upon us?

The study of the placebo effect in Parkinson's disease solves one part of the puzzle—namely, the way in which belief alters the brain. Simply because people expected something to happen—in the case of Parkinson's disease, an amelioration of their motor problems—it often did, resulting in the release of dopamine. This is the clearest demonstration to date of the power of faith and the optimism it inspires to alter brain chemistry. Granted, in terms of the experiments that have been conducted, the faith exhibited was in the form of a pill rather than belief in God, but the general principle is the same.

That leaves a second mystery: How does altered brain chemistry translate into selective advantage? The association of religiousness and prayer with longer life may partially answer this riddle. After all, one has to be alive to reproduce. The effects of placebos and nocebos point in the same direction. People who believe they will be healthy have a better chance of surviving than those who expect to die early. Genes that promote such beliefs are more likely to be replicated and handed down to the next generation. But such effects by themselves are unlikely to account for the persistence of God genes over the long stretch of human evolution. For one thing, as noted earlier, they are mostly evident at the end of life, long after the period of maximum reproduction. Something more is required.

Genes are selfish. The only thing they care about is being passed on to the next generation. Since the only way that can happen is through reproduction, the essential point to consider about the evolution of God genes, or genes that help to promulgate one's need to believe, is how they contribute to procreation. How might a gene such as VMAT2, which encodes a monoamine transporter, increase the rate of human reproduction?

Part of the answer may be psychological. Dopamine, one of the several monoamines that are influenced by VMAT2, is the brain's reward chemical. It is what makes people happy, confident, and optimistic. Somebody who feels good about the future may be more likely to get up and hunt for food, build a shelter, and—above all—want to have children.

On the other hand, serotonin, another monoamine that is packaged by VMAT2, is the brain's "feel-bad" chemical. Shifts in serotonin levels can cause people to feel lonely and worried about the future. Although it might seem counterintuitive, this might actually increase reproduction among those people who desire children as objects of love and affection, and, in some cultures, as a source of support in old age.

Besides these psychological mechanisms, VMAT2 may have more direct effects. Both dopamine and serotonin are known to significantly influence reproductive behavior. Dopamine plays a role in partner diversity through its effects on novelty-seeking behavior. For example, people with a particular variation of one dopamine transporter gene were found to have more sexual partners than those with another version of the same gene. It's not difficult to imagine how this could have resulted in more offspring, especially in the early days of human evolution.

Serotonin influences how often people have sex. This is the reason that many serotonin-altering antidepressant drugs have the unpleasant side effect of reducing libido and sexual functioning. It is also why people with one version of the serotonin transporter gene have sex more frequently than do those with a slightly different genotype. While we are not sure whether more serotonin or less causes low libido, we do know that the shift in serotonin that causes depression (either an increase or a decrease—scientists aren't sure) is the same shift that is associated with a higher libido. Again, it is likely that such differences could translate into differences in reproductive rate.

Of course, these are only speculations, as are so many theories in evolutionary psychology and sociobiology. To prove them, we would need to replay the tape of human evolution with and without these "God genes." The important point is that a biochemical and neurological understanding of how spirituality works in the brain gives us a basis for making reasonable proposals. As we learn more about these processes, our guesses can only become more educated.

Nine

Religion

From Genes to Memes

A great many people think they are thinking when they are merely rearranging their prejudices.

—*William James*

One's spirituality is a private affair, a matter of our inner feelings and beliefs. But spirituality does not exist in a vacuum. It almost invariably becomes entwined with its more public, structured form: religion.

The question of the origins and evolution of religion has puzzled philosophers, theologians, and historians for centuries. Obviously, I am not equipped to answer how religion developed here; after all, I am just a behavior geneticist. Nevertheless, I believe that certain insights from biology and psychology can help to see religion in a new and useful light. One of the most useful recent concepts that offers some insight into religion and theology was invented by a geneticist, Richard Dawkins. It's called the meme.

The Meme Machine

Memes, according to Dawkins, are transmittable units of culture. Songs, poems, and advertising jingles are examples of memes. So are clothes, fashions, and cosmetic surgery. Memes can be methods of communication such as e-mail, mathematical techniques such as long division, and technological conveniences such as speed dialing. There are small memes, such as the idea that every room should have a touch of yellow in its decoration, and large memes, such as democracy.

The critical feature of memes is their ability to be replicated. Indeed, Dawkins deliberately chose the name "meme" to sound like "gene" because both of them can be faithfully copied. The difference is that genes use cellular enzymes or viruses, whereas memes use imitation. Susan Blackmore defines memes as "instructions for carrying out behavior, stored in brains (or other objects) and passed on by imitation" in her lucid book *The Meme Machine* (itself a successful meme, judging from the subtitle of this section).

Memes are like genes in another way. They are selfish. They only care whether they are copied or not, not what happens to the copier. This is why memes such as "smoking cigarettes is cool" retain their popularity despite the harm they do. As long as the meme is efficiently copied from one brain to another—a process aided by advertising—it will persist.

Memes may be as useful for understanding the transmission of culture as genes are for understanding biology. Despite the many similarities between them, however, there are also important differences.

First, genes are found in all living creatures, whereas memes largely are limited to humans. True, there are a few examples of culturally learned traits in animals: female guppies pick the same mates as their peers, British swallows open milk bottles, and certain chimpanzees crack nuts with stones or fish for termites with twigs. (A much longer list of examples is given in Lee Dugatkin's delightful book *I'll Have What She Just Had.*) But these are the exception, not the rule, and fall far short of the complexity and richness of human memes.

Part of the reason for the difference is that humans have larger brains than animals, allowing them to store more information and process it more quickly. An even more important distinction, however, has to do with our communication skills. We have language, animals don't. That makes it a lot easier to copy a new idea.

A second important difference between genes and memes lies in the efficiency of reproduction. Genes can't be copied any more rapidly or productively than the geometric replication of DNA. One copy makes two, two copies make four, and so on down the line. Because it is impossible for a human female to have any more than about 20 children in her lifetime, she can pass on at most 20 copies of her genes; some powerful or attractive males may do a bit better than that, but most will do worse.

Memes do not have this limitation. They can be passed on just as rapidly as they can be communicated. One 30-second advertisement on Super Bowl Sunday can make more meme copies than the entire world's population makes gene copies in a year. Here the formula is one copy makes 50 million.

The third difference is in time frame. Genetic evolution is slow. Most of our DNA sequences have been with us for millions of years, and even the most rapidly changing genes—such as those for viral resistance—have taken centuries to evolve. But memes can change overnight. Think of hula hoops, Studio 54, business guru Armand Hammer, and artificial granite countertops. (If you have no idea what I am talking about, that just proves the point.)

Because of these unusual properties, there are few logical constraints on the reproduction of memes. It makes no difference if they are true or false, harmful or helpful, short-lived or long-lasting. If a meme can efficiently colonize the human brain, it will. So what does this have to do with religion and faith? Consider two examples.

St. Francis's Prayer

In 1587, a small fishing boat working just off Calabria on the southwest coast of Italy was hit by an unusually violent storm that threatened to capsize the boat and kill everybody aboard it. The captain, not usually a pious man, had shortly beforehand visited an order of Franciscan friars led by a certain Father Francis, a local fellow who had recently built a chapel and monastery overlooking the very sea on which his ship was now foundering. The captain prayed to God and to Francis for deliverance.

His prayers were answered. The storm quieted down, and the ship made it safely to shore. Ever since, Catholics have honored St. Francis of Paola as the patron saint of sailors at sea and all other voyagers. His likeness can be found in taxicabs all over the world.

The prayer to St. Francis is a classic example of a meme. It owes its continuing popularity to several factors.

The first is its irrefutability. The original sailor told other people that his prayer had worked—when his ship was in danger of sinking, he prayed to God and was saved. But suppose that he had prayed to God but had not been saved. In that case, he would have died, and nobody would have heard about the fact that his prayer hadn't worked. Or suppose he had not prayed at all, but was saved anyway. Again, people would not have attached any special significance to it, since no prayer was offered. In other words, the meme for prayer only gets replicated if a prayer is made and it seems to work. It has a reasonable chance of being believed and little chance of being disproved.

The second factor regarding prayer is that it's cheap in terms of time, money, and effort—one can pray to St. Francis in the same amount of time it would take to sit and fret about the situation. As Pascal noted long ago, this type of investment has little to lose and much to gain.

The third aspect about prayer is that it feels good—regardless of whether our prayers work or not. As noted in the previous chapter, positive expectations can actually lead to the release of dopamine, the brain's "feel-good" chemical, in Parkinson's disease patients. The same mechanism likely works just as well for desperate sailors, anxious taxicab drivers, and others who believe.

This isn't to say that prayer to St. Francis does not work. It very well may. The point is that, given the above, people would continue to practice and pass on this meme even if it did not.

The Circumcision Meme

Each male among you must be circumcised; the flesh of his foreskin must be cut off.
This will be a sign that you and they have accepted this covenant.
Every male child must be circumcised on the eighth day after his birth.

—*Genesis 18:11*

Infant circumcision has been practiced by Jews for more than 3,000 years. What is the appeal of this ancient ritual?

Some have suggested that the main advantage of circumcision is medical. In modern times, circumcision has indeed been shown to reduce urinary tract infections, venereal disease, and penile cancer. But when this practice began, any such advantages must have been far outweighed by the risks of surgery with an unsterilized knife. Having the *mohel* suck the blood from the circumcised penis with his mouth, a practice introduced in the Talmudic era, further increased the hazard of the operation.

The real advantage of circumcision is that it's a marker for being a Jew. The tradition started, according to the Hebrew Bible, when God told Abraham that every male Israelite should undergo circumcision as a symbol of the covenant between God and his chosen people. Ever since, circumcision has been a way for Jews to tell the difference between friends and enemies. The early history of Judaism is one of constant struggle with surrounding tribes and clans, most of whom were physically indistinguishable from one another. A circumcised penis was a handy sign of who was on your side.

The importance of the sign of circumcision is highlighted by two of the most disquieting stories in the Bible: the rape of Dinah and God's attempt to kill Moses.

According to the book of Genesis, Dinah was a Jewish woman who moved with her family to the land of Canaan—the West Bank of its day. One day, on her way to see friends, she was raped by Shechem, the son of the local Canaanite chieftain. Shechem fell in love with his victim and asked her family if he could marry her. He also proposed that the Jews and Canaanites intermarry and live together in peace.

Dinah's brothers guilefully agreed, but only under one condition: Shechem would have to be circumcised, and so would all his people. They couldn't possibly let a Jewish woman marry an uncircumcised man, they said; it would be sinful. If the Canaanites wanted to share genes with the Jews, they would first have to share their memes. The implication was that being uncircumcised was even worse than being a rapist.

(As it turns out, the demand for circumcision was a clever ruse. Once Shechem and his kinsmen had undergone the painful and temporarily crippling operation, Dinah's brothers killed every last one of them.)

The importance of circumcision is even more dramatically highlighted by the biblical story of God's attempt to kill Moses.

> On the journey, when Moses and his family had stopped for the night, the Lord confronted Moses and was about to kill him. But Zipporah, his wife, took a flint knife and circumcised her son. She threw the foreskin at Moses' feet and said, "What a blood-smeared bridegroom you are to me!" After that, the Lord left him alone. (Exodus 4:24–26)

Most religious scholars believe that God was angry at Moses because he failed to circumcise his son. The implication is that circumcision was a very important custom or belief. It takes a lot to get God mad enough to kill.

Why was having a way to recognize other Jews so critical? Genes have an increased chance of persisting when closely related kin, who are likely to carry the same genes, help one another out. It's usually easy to distinguish kin from unrelated people by visible genetic traits such as eye color, skin hue, and facial appearance.

Memes, on the other hand, have an increased chance of survival when individuals who share beliefs cooperate with one another. Since cultural kin are not biologically distinct, they must invent and incorporate signs of belonging into the fabric of the cultural group. The ideal mimetic marker is one that is at once readily apparent, so that group members can readily recognize one another, yet at the same time easily camouflaged, so that one can escape persecution by cultural outsiders. Circumcision is a perfect solution.

The use of mimetic signs to mark identity, such as circumcision, is common in religion. Abstaining from eating beef or pork, wearing yellow robes or a turban, shaving one's head or shaking a tambourine at the airport—all of these say to others with the same beliefs, "I belong with you."

Are memes purely cultural, or are they influenced by genes and biology? To find out, scientists turned once again to the study of identical twins.

Tony and Roger

The identical twins who later became known as Tony Milasi and Roger Brooks were born five minutes apart on May 28, 1938, at the City Hospital in Binghamton, New York. Their mother was Catholic, and their father was Jewish. Because the parents were married to other people, they decided to give the children up for adoption shortly after birth. As an unwitting consequence, the twins received such different upbringings that they became a perfect test case for the effects of nature versus nurture on religion.

Within two weeks, Tony was adopted by a local Italian American couple, Pauline and Joseph Milasi, who had been trying to have a child for many years. They were so happy to have a son that they treated him as if he were a reincarnation of his namesake, St. Anthony.

Tony's adoptive mother, Pauline, was a pious woman. At age 46, she had suffered through the disappointment of three miscarriages, two stillbirths, and two children who were born healthy but died in infancy. Prayer to St. Anthony of Padua was her last remaining hope for a child. Tony was the answer to her supplications.

Pauline impressed Tony with fear of the Lord from an early age. She taught him his prayers, first in Italian and then in English, in front of the altar to St. Anthony that she had made in her bedroom. The room was lit by devotional candles on the dresser.

"I was taught to pray hard," Tony later recalled in a book that was written about the twins. "Some nights . . . I'd be so scared that I'd promise in my prayers not to swear or talk fresh and that I wasn't going to put my hands on this neighbor girl's breasts anymore."

All of Tony's friends and schoolmates were Catholic, and they went through their religious training and rites of passage together. A photograph shows him smiling angelically in his white confirmation suit. Not surprisingly, he became an altar boy at his local church at the age of 12. He enjoyed the job and made sure to be at the early-morning service even when it was snowing outside.

The parish priest told Pauline that "God was well pleased" with her son. Although Tony didn't understand the words to the mass, which was said in Latin, he enjoyed singing the responses to the priest, carrying the Bible, and ringing the bells at the consecration. The ritual pleased him.

Meanwhile, Roger was experiencing a more secular upbringing. His first caretakers were a heavy-drinking, unmarried couple who allowed him to be badly burned in a hotel-room fire. After that he went through a series of orphanages, foster homes, guardians, and foster parents, living in nine different homes by the age of ten. He subsequently was taken in by Mildred Brooks, the wife of his biological father, who felt guilty for her husband's indiscretions.

Although Mildred was nominally Jewish, she was a nonbeliever who rejected all religions. She never made any effort to take Roger to synagogue or provide him with Jewish religious training. The only one who did was her mother, Bessie. But even she dismissed any thought of serious religious training for Roger with the comment that "he doesn't have a Yiddish *kopf* [head]."

Although most of Roger's peers were non-Jewish, he did become friends with Ray Skop, the son of a rabbi, and soon was spending every Friday night at their house. Roger was captivated by the Shabbat ceremony. He loved to don his yarmulke and join in the prayers and songs. "For a young boy he had a deep sense of religiosity," remembered Rabbi Skop. "He was fascinated by the *Zmirot,* our table songs that we chant after the meal."

That was as far as Roger progressed in religious training. He dropped out of Hebrew school after only a few classes and was never bar mitzvahed.

The story of Tony and Roger illustrates two key points about the roles of environment and heredity in religion and the difference between memes and genes. First, *what* people believe in is purely learned. Even though they had the same DNA, Tony and Roger grew up with and held very different religious beliefs. Tony accepted that Jesus is the son of God, that the Pope is infallible, and that it's a sin to use birth control. Roger disagreed with each one of these doctrines. Religion, in other words, is part of what we absorb from the culture we grow up in.

Second, *whether* people believe, and how deeply, may be influenced by their genetic makeup. Neither Tony nor Roger was a particularly spiritual individual, yet both of them became interested in religious ritual at a similar time in their lives. Tony became fascinated with Catholicism and Roger with Judaism, perhaps because of the sense of importance that religion gave to each of them.

Tony told researchers he liked being an altar boy because it made him the focus of attention. He looked good in his black cassock and white surplice. Everybody was watching him. He didn't pretend to be deeply spiritual and never claimed to have a personal religious experience. He was more interested in winning the Altar Boy of the Year Award, which would give him a week's free vacation at a Catholic camp.

Roger's interest in Jewish ritual, too, was more social than spiritual. Talking about his Friday nights at Rabbi Skop's house, he later remembered, "After our Shabbas meal we all walked to temple. I got to sit in the front row and I felt important." Going to the Skops' had a certain prestige associated with it. "To my grandmother," Roger said, "overnight at the rabbi's was better than if I got invited to the White House."

The story of Tony and Roger reinforces that the content of religious ideas and traditions is cultural, whereas the predisposition to believe them may be at least partially genetic. But, of course, I've based this anecdotal evidence on but one pair of twins. Do such preliminary findings hold up in the population at large? To find out the answer to exactly that question, scientists turned to large-scale studies.

Religious Affiliation

Do people become Catholic or Muslim, Mormon or Baptist because they were raised that way or because of their genes? Certainly a person's religious affiliation is learned, isn't it?

To test that assumption, scientists in Australia analyzed data from 705 pairs of twins who were living together with their parents, and another 3,025 pairs who had left home and were separated from one another. All of the twins and their parents filled out questionnaires. They were asked to categorize themselves as Anglican (the largest religious group in Australia), other Protestant, Catholic, Jewish, or Greek or Russian Orthodox. Once the data was collected, the numbers were fitted to mathematical models that distinguished various types of cultural influences from genetic inheritance. By obtaining information about the parents as well as the twins, it was possible to infer not only whether religious affiliation was due to genes or environment, but also how it was passed on.

As one might predict, the identical twins and fraternal twins raised and still living together overwhelmingly shared their religious affiliation with each other. The correlation was 88 percent for the identical, monozygotic pairs, and 86 percent for the fraternal, dizygotic pairs. Clearly, it didn't make any difference how genetically similar the twins were. Some other mechanism was involved in their attraction to the same religion.

For the twins still living at home, the answer was obvious: their parents. This is expected, since parents are the ones who guide their children's early religious training. Kids usually don't just wake up one morning and decide to attend a church or synagogue or mosque; they go where their parents take them. It is natural for parents to pass on their faith and religious convictions to their children.

The study, however, suggested that two interesting changes occurred in the religious beliefs of the twins as they grew up and separated from each other and their parents. First, although cultural background remained a strong influence, the role of parents diminished in the twins relative to other shared environmental factors. The twins still usually had the same religion, but as they grew older it was more often different from that of their parents. In other words, once the twins were out of their parents' house, they were more likely to be swayed by new ideas, new trends, new friends, or each other than by Mom and Dad.

Perhaps, for example, one twin was "born again" and was so moved by the experience that she convinced her twin to join the same denomination or church. Or perhaps both twins discovered a religion that didn't exist in the neighborhood where they grew up. Both of these represent nonparental forms of cultural transmission.

But a second change was noticed as well, an unexpected one. Genes seemed to play a role in their beliefs. The influence was weak—it was statistically significant only for females, rather than males, and was still a less powerful influence than the twins' shared environment. But it was there. Does this mean that there really are "Catholic genes" or "Muslim genes"? Probably not. The more likely explanation is that once the twins were off on their own, their decision as to whether to maintain or reject the religion of their parents was indirectly influenced by genetically mediated personality traits.

Let me recount the story one of the most famous religious converts of the twentieth century—Edith Stein, the "Jewish saint." Stein was born in Poland on Yom Kippur, the Jewish Day of Atonement, in 1891. Her parents were Orthodox Jews and her great-grandfather had been an Orthodox cantor. Despite a traditional upbringing, Stein was a rebel from early on. At the age of 14, she declared herself an atheist. In an era when higher education was unusual for women, she insisted on attending university and became an eminent philosopher. Long before the women's movement, she was a feminist who believed that women should have the right to vote and have independent careers.

It really is not so surprising, then, that Stein had an overnight conversion to Catholicism at the age of 29 after reading an autobiography of Saint Teresa. After becoming a Carmelite nun, she continued to rebel, refusing to leave her monastery for a safer spot when her life was threatened by the Nazis because of her Jewish origins.

Stein's life ended in 1942 in a gas chamber at Auschwitz. She was beatified as Sister Teresa Benedicta in 1987 and canonized in 1998—the first Jewish woman since Mary to be sainted.

The point of Stein's story is that her conversion from Judaism to

Catholicism was neither a fluke coincidence nor a direct result of any particular set of environmental circumstances. It was part and parcel of her self-reliant character. If she'd had an identical twin, her twin probably would have been just as self-reliant, and therefore just as likely to change her religion. It wasn't Catholicism that was in Stein's genes; it was a questioning character that predisposed her to radically throw off her past and embrace a new tenet of beliefs.

Orthodoxy

Although the content of religious memes is clearly determined by culture rather than biology, the extent to which people embrace religious traditions and beliefs could very well be influenced by their personality—and therefore by their genes. To find out to what extent this actually occurs, scientists asked the same large group of Australian twins who participated in the study of religious affiliation questions about their specific religious beliefs—for example, whether they regarded the Bible as literal truth.

In analyzing the results, three possible sources of variation were considered. The first was the shared environment, which included everything the twins learned from being brought up together. If the twins were taught by their parents that God created the universe in seven days, that would count as a component of their shared environment. The second source of variation was genetic. If the twins accepted that the universe was created in seven days because they were genetically predisposed toward belief in a higher power, that would weigh in as a genetic contribution. Finally, there was the nonshared environment—all influences outside of genes and shared environment. If one twin had a schoolteacher who clearly explained evolution, thereby leading that twin to reject creationism, that would count as part of the twin's nonshared environment. So, too, would separate friends and unique experiences.

The results from the twin study showed that both culture and genes play a role in acceptance of religious beliefs. As might be expected, there was some quantitative variation depending on the particular belief. Those who hold that the Bible was absolute truth, for example, were influenced by culture to a considerable extent (34 percent), while genes played a smaller role (25 percent). Belief in divine law was about equally split between genes (22 percent) and environment (26 percent), whereas whether to observe the Sabbath was tilted toward heredity (35 percent), with shared environment less important (18 percent). The remainder of the variation for each trait could be ascribed to that mix of life events, serendipity, and measurement uncertainty called unique environment.

Factor analysis showed that these orthodox religious beliefs were tied to attitudes about other social and religious issues, such as evolution, divorce, birth control, and abortion. When a religious orthodoxy factor was made by combining all of the above beliefs, scientists found that the twins' shared environment was the biggest factor (at 40 percent), while one's genes, too, played a significant role (at 27 percent).

Research in the United States has confirmed the role of both nature and nurture in the acceptance of traditional religious beliefs. A study of 820 adolescent and young adult twins from Indiana found that a religious orthodoxy scale, which measured adherence to doctrine and acceptance of traditional beliefs, was influenced 61 percent by cultural learning and 10 percent by genes.

The important lesson from these studies is not the precise percentages ascribed to genetic differences or to one's shared environment, but rather that both play a significant role. One reason, perhaps, that particular religious beliefs are so commonly accepted is that they connect with natural properties of human brains—our desire for certainty, for example, or our desire to feel good. And perhaps the reason that some people embrace such beliefs more than others is genetic differences in how those beliefs are received by the brain and how they make those people feel.

Like Father, Not Like Son

The twin studies confirm that religious beliefs are learned. But where are they learned from? Most parents would like to think their children learn their beliefs from them. The availability of large twin data sets, including parents, siblings, children, and spouses, now allows that notion to be tested. By comparing family members, it is possible to determine from whom twins' religious attitudes and behaviors are learned, and how much they learn.

The results are surprising. As it turns out, parents have little impact on their children's beliefs. And when they do, their impact is often the opposite of what they would have hoped.

The results of a large study in Virginia, for example, show that parental influence accounted for only about 1 percent of the variation in a religious orthodoxy scale among twins that included such questions as "support for school prayer." Moreover, the little impact that parents did have was in most cases negative; that is, the children of parents who did believe in school prayer were less likely to support this idea themselves. This was true for the relationships between fathers and daughters, fathers and sons, and mothers and sons. The only instance in which parents had a positive association was between mothers and daughters.

Religious attendance reveals a similar pattern. In the Virginia sample, parental guidance had a less than 2 percent effect on how often their children went to church, and the effect was actually negative for the influence of fathers on sons. The results from an independent study in Australia show that parents down under had a similarly meager impact.

Why do parents seemingly have so little effect on their children's beliefs?

Rebellion may be one answer. Bucking one's parents is a normal part of growing up. It's an important step in establishing an individual identity. Dragging kids to church every Sunday is as likely to make them dislike it as to make them love it. Similarly, telling them what to think about controversial issues like abortion and school prayer may just turn them in the opposite direction.

Trying to keep them in the dark about controversial topics will not work either, owing to the ready access to information that is provided by the modern press, other media, and the Internet. The take-home message for parents is that telling a child what to do or how to think may encourage him or her to behave in exactly the opposite manner.

The Power of Belief

Many religious beliefs, such as the concept of continuing consciousness following death, have been with humans since before recorded history. One of the most important factors behind the power and persistence of such ideas is their cohesion, which is achieved in two ways.

First, most religious beliefs are mutually supportive of one another. For example, the belief that "God listens to your prayers" also supports the belief that "God hears your curses." Similarly, the injunction to "treat others as you would have them treat you" supports the teaching that you should "turn the other cheek." The rite of circumcision goes hand in hand with the belief that "we are the chosen people" and thus need to be identifiable in certain similar ways.

The second factor behind religious cohesion is the tendency—in some religions, almost a requirement—that one marry and have children with people of one's own faith. This belief has been remarkably well obeyed over the course of Jewish history—so well that the same DNA markers found in the original Jews are still present in the Jewish population today. People have a strong tendency to choose spouses with the same faith, the same beliefs, and even the same level of religious attendance. The result is to reinforce both the cultural and genetic underpinnings of religiousness.

Some of the clearest data to support this comes from the same twin family studies used to analyze how religious beliefs and ideas are transmitted. In the Virginia study, for example, the correlations between husbands and wives come in at 71 percent for attendance at religious services, 68 percent for religious affiliation, and 45 percent for religious orthodoxy. The data on Australian spouses was similar; the correlations were 74 percent for church attendance and 72 percent for denomination.

What these numbers suggest is that most men and women marry partners of the same faith or beliefs. They usually go to the same church (or don't belong to any). For the most part, they have the same religious beliefs and the same attitudes about issues like divorce and abortion. In short, they are very much birds of the same religious feather.

This assortative mating, as geneticists call it, leads to a strong clustering of religiousness in families. Most personality characteristics—thrill seeking and depression, for example—are shared only about 10 to 30 percent by nuclear family members. But the correlation between religious beliefs and behaviors is often 50 to more than 70 percent—the sort of similarity that usually is seen only in identical twins. When parents have the same religious affiliation, beliefs, and practices, they are all the more likely to pass them on to their children by both teaching and heredity. The kids get double exposure, from both their environment and their genetic heritage. That's why—to reverse the old adage—the family that stays together prays together.

Ten

The DNA of the Jews

Truth for us is simply a collective name for verification processes.
—*William James*

The Sinai Desert is a harsh place. It is hot in the day, cold at night, and arid all the time. There is no green to relieve the unremitting red and brown of rock and sand. Roughly in the middle of the peninsula, halfway between Eilat to the north and Sharm el Sheik to the south, the Gulf of Aqaba to the east and the Gulf of Suez to the west, stands Jabal Musa, a craggy peak that rises 6,855 feet above sea level. It can be seen from many miles away.

According to the Bible, this is where Moses led the people of Israel some 3,000 years ago. They made their camp at the base of Mount Sinai, arguably the present-day Jabal Musa, two months after the flight from Egypt. While they were there, God made them an offer: "If you will worship me and obey the Ten Commandments and Torah, I will lead you out of the wilderness to the promised land." The people accepted. They didn't have much choice. Behind them was a bunch of angry Egyptians; ahead lay a barren wasteland.

Shortly thereafter, God instructed Moses, his brother, Aaron, Aaron's sons, Nadab and Abihu, and 70 of the Israelite leaders to climb the mountain. Halfway up, they encountered God himself, a terrifying presence who, standing on a platform of brilliant clear sapphire, commanded Moses to continue his ascent alone while the rest remain behind. Moses climbed through the clouds to the top of the mountain, where he spent the next 40 days and 40 nights receiving written and oral instructions from God, including this edict concerning the priesthood:

> Your brother, Aaron, and his sons, Nadab, Abihu, Eleazar, and Ithamar, will be set apart from the common people. They will be my priests and will minister to me.

According to the Bible, God then gave detailed directions on the ceremony for the ordination into the priesthood of Aaron and his descendants, who were to become the Jewish priestly caste known as the Cohanim. God described the animals to be sacrificed, the food to be eaten, and even the precise clothing to be worn.

Little did Moses and the Israelites know that some three millennia later, it would be possible to check how accurately they had followed God's instructions by a new technology: DNA testing.

DNA As History

DNA carries two different types of cellular information, functional and historical. Most of this book has been about the first type of information, which consists of instructions on what types of proteins and RNA molecules to make in different cells and tissues at various times in development. This is the way that genotype influences phenotype—whether it is through a pigment gene that causes blue or brown eye color, or a God gene such as VMAT2 that leads to lower or higher levels of self-transcendence.

But DNA also carries historical information. DNA is the product of evolution and reproduction, processes that connect each organism to the preceding generation. By looking carefully at a person's DNA, it is possible to tell who that person's ancestors were—both in the immediate sense of parents, grandparents, and so on, and in the long-term sense of discovering our biological ancestors and cousins: chimpanzees, gorillas, and other creatures.

In this chapter, I want to explore the use of DNA as a historical record of one particular religious tradition: Judaism. Although the DNA sequences used for this purpose do not have functional significance—they don't even code for proteins—they can be used as molecular markers for an individual's genetic heritage. By studying the worldwide distribution of such sequences, it has been possible to trace the migrations of the original Israelites, to test their adherence to God's reported commandments about the inheritance of the priesthood, and to infer the role of certain cultural practices as genetic selective forces.

The evidence comes from studying the Y, or male, sex chromosome. Men have one copy of the Y chromosome and one copy of the X chromosome, whereas females have two X chromosomes. Because men inherit their Y chromosome only from their fathers, it is a good marker for studying paternal lineages—a sort of genetic surname. And because the Y chromosome is passed on without any reshuffling of the genetic information, it is a particularly useful tool to reconstruct population history.

All present Y chromosomes can be traced back to one progenitor who lived in Africa about 140,000 years ago. Scientists call him Adam. He was probably not the only human being at that time, but he is the only one who begot sons at every subsequent generation; the others must have had a descendant who had no children or only daughters, which would cause their paternal lineage to die out.

If DNA were perfectly copied at each generation, every human would have precisely the same Y-chromosome DNA as Adam. But the DNA duplicating machinery is not error-free. Mistakes get made. Occasionally, one of the building blocks of DNA gets replaced or left out completely. This results in differences in DNA that can be used as markers to distinguish one male, and his descendants, from another. This is the basis for paternity testing. For example, it was Y-chromosome markers that were used to show that Thomas Jefferson (or his brother) had children with his slave Sally Hemings.

Y-chromosome DNA markers also can be used to study the history of groups of people. When Y chromosomes are logically arranged according to the number of differences they have compared to Adam's DNA, they form a branched tree. Closely related people occupy nearby twigs. People who have been separated for a long time are located on more remote branches. By measuring the precise distance between branches, it is possible to estimate how long one group has been isolated from another. Such trees are ideal for studying the history of populations as they divide, migrate, and merge with other peoples—processes that the Jewish people are particularly familiar with.

The Wandering of the Jews

At the time of Moses, the Jews consisted of twelve tribes, or family groups, of nomadic Middle Eastern Semites. God had promised to settle them in the land "all the way from the border of Egypt to the great Euphrates River" if they would honor and obey him, according to the Bible. This proved to be a difficult contract for both sides to honor.

First, the Israelites refused to move into Canaan because they feared its occupants. God retaliated for their lack of faith by making them wander in the wilderness for 40 years before they finally entered the promised land through the walls of Jericho.

No sooner did the Israelites arrive than they started fighting among themselves. This led to the separation of the promised land into two kingdoms: the ten tribes of Israel in the north and the two tribes of Judah in the south. In 721 B.C., the north was taken over by the Assyrians, and its people lost their identity as they were scattered to distant lands and assimilated by the locals. A century later, the two tribes in the south were defeated by Nebuchadnezzar and packed off to Babylonia. Most but not all of these Jews, as they were beginning to be known, returned to Judah 60 years later when the Babylonians were defeated by the Persians. There they remained for the next five hundred years, which is known as the Second Temple period, ruled first by the Greeks, then by the Seleucids, and then by themselves.

Then the Romans arrived. By A.D. 70 they had taken over Jerusalem, burned the temple, and sent most of its inhabitants off to slavery. There was not to be another official Jewish state for almost two millennia. Following a second revolt in A.D. 135, the Romans became even more vengeful. They razed the entire city of Jerusalem, outlawed the religion, killed the Jewish elders and leaders, and renamed the area Palestine.

The Jews again became nomads. By the first century A.D. they had spread out around the Mediterranean to form Jewish settlements in Alexandria, Carthage, Pompeii, Athens, Cyprus, and Crete. There were Jews who ended up in Sinope on the Black Sea, in Babylon on the Euphrates, and in Ctesiphon on the Tigris. Many Jews were sent as slaves to Rome, whence they were dispersed to every corner of the empire. Strabo, a Greek geographer, wrote, "It is not easy to find any place in the habitable world which has not received this nation."

The Middle Ages were another difficult age for the Jews; Christians considered them to be the "killers of Christ." As the crusaders swept through Europe, they drove the Jews to the north and east, into modern-day Germany, Poland, Lithuania, and Russia. In Spain, the large Jewish population was given the choice of conversion or death. Ghettos and severe work restrictions for Jews were introduced. Shakespeare's "Merchant of Venice" was a money changer not by choice but because Jews were allowed few other occupations.

Twice, the entire Eastern European Jewish population sank to fewer than 50,000 people: in 1350 during the plague, and in 1700 following a particularly vehement series of Cossack massacres. These biological "bottlenecks," where the population declines precipitously, are the reason that certain genetic diseases, such as Tay-Sachs disease and torsion dystonia, are so common among Jews of European descent today. If the disease-causing mutations happened to be present in one or more of the people who passed through the population restriction, their frequency in subsequent generations would be elevated.

In the middle of the twentieth century, Nazi Germany would cause another devastation on the Jewish population: the Holocaust. In just five years, 6 million Jewish men, women, and children were killed.

As the Jews spread to every corner of the globe, they adopted the customs, dress, and behavior of each country's native populations. Slowly but surely they began to resemble, at least superficially, the indigenous populations into whose midst they were thrown. This eventually led to the separation into two major Jewish groups, the Ashkenazim and the Sephardim, whose appearances are as distinct as Europeans and Arabs.

By the time of the reestablishment of a Jewish state in 1948—

what Israelis call the War of Independence (and Palestinians know as The Catastrophe)—Jews were a far more heterogeneous people than they were when they left that region some 2,000 years earlier. Walking the streets of Jerusalem or Tel Aviv or Beersheba today, one sees Jewish people of all sizes and shapes and colors: Jews with red hair, green eyes, and freckles; Jews with blond hair and blue eyes and skin so fair that it sunburns at the least provocation; Jews with black hair and brown eyes and skin shades ranging from olive to eggplant; Jews with round eyes and Jews with almond-shaped eyes; Jews who would look at home on the streets of New York, Cairo, St. Petersburg, Cape Town, or Rome—which may be exactly where they are from. All of these modern-day Israelites are Jewish, and about one in twenty identifies him- or herself as a Cohen. Are they really a people in the genetic sense?

Aaron's DNA

It took a Cohen, Karl Skorecki, to answer that question.

Although the Cohanim lineage has tried to maintain its traditions, it hasn't always been easy. Their main function among early Jews—to offer animal sacrifices to God—was eliminated when the second temple was destroyed by the Romans in A.D. 70. And their roles as teachers and legal authorities were taken over by the rabbis during the Diaspora. These days, a Cohen has little power or privilege within the Jewish community besides the five silver coins he receives for redeeming a firstborn son from service to the Temple—a gift that usually is returned to the child.

The Cohanim do, however, continue to pass on the priesthood from father to son, to bless the congregation on festivals, and to have first dibs at reading the Torah in synagogue. One day, Karl Skorecki, who is of Czechoslovakian descent, was at synagogue in

Toronto when a fellow Cohen from North Africa was called to perform the reading. Remembering God's words in the Bible, Skorecki wondered whether he and the dark-skinned African man were genetically related. Being a physician, he immediately thought of DNA testing to find out the answer. Soon he was on the phone with Michael Hammer, an American scientist who developed some of the first Y-chromosome markers, and Neil Bradman, a Jewish scientist in Great Britain who had already started collecting DNA samples from Cohanim and other Jews.

Together they compared two different Y-chromosome DNA markers in 68 Cohanim and 120 lay Jews from Israel, Britain, and North America. The results were intriguing. There was one particular pattern that was present in a substantial portion of the laymen but almost completely absent in the priests. It was the first hint of a difference between the two groups at the DNA level.

But these initial results were less than conclusive. One problem was that although the scientists had discovered a pattern that was mostly missing from the Cohanim, they had not found one that was usually present, which would be far more convincing. Another was that data did not show when the difference between Cohanim and lay Jews arose. If the biblical tradition was true, the separation should have taken place about 3,000 years ago; but if the divergence was more recent, it might have resulted from an entirely different cause. For example, people didn't start to use last names until the Middle Ages. If the DNA difference between the Cohanim and laymen dated to this period, it might represent nothing more than a signature for a particular set of surnames that individuals with priestly functions happened to adopt at that time. (Such a DNA signature for a surname has actually been demonstrated for the last name "Sykes," which first appeared in England about seven hundred years ago.)

To resolve these uncertainties, the scientists needed to look at more DNA markers and to have some way of dating their divergence. They called in David Goldstein, a British scientist who pioneered the use of the Y chromosome as a molecular clock.

The new study used twelve Y-chromosome DNA markers rather than two, an important improvement. Half of the markers were DNA sequences that change very slowly, probably only once in the history of man since Adam. These served as the hour hand in the molecular clock, useful for judging when ancient divisions between people occurred. The rest of the markers were located in more unstable bits of DNA that sometimes change in a single generation. They would be used as the clock's second hand. The study included 306 unrelated men—119 lay Jews, 106 Cohanim, and 81 Levites (who represent a sort of junior priesthood held together by different sets of religious rules). About half of all those studied were Ashkenazic; the rest were Sephardic.

This time the scientists found what they were looking for: a distinctive genetic signature for the Cohanim. The DNA pattern was so clear that the scientists even gave it a name: the CMH, or Cohen modal haplotype. This pattern was present most often in Cohanim, less frequently in lay Jews and Levites, and hardly at all in most non-Jewish populations.

Importantly, the same unique DNA sequences were found in both Sephardic and Ashkenazic Cohanim, meaning they have to do with being a Jewish priest, not whether they were a European or Arab. Forty-five percent of those who were Ashkenazic Cohanim had the CMH marker, 56 percent of those who were Sephardic Cohanim had it, 13 percent of those who were lay Ashkenazim had CMH in their genetic signature, and 10 percent had it who were lay Sephardim. When small variations that might have been due to recent DNA changes were discounted, the difference between priests and laymen became even more pronounced. And when only the most stable DNA markers were analyzed, more than 95 percent of Ashkenazic and 87 percent of Sephardic Cohanim had the same simplified haplotype.

The next question was one of timing. Did the separation between Cohanim and other Jews occur in ancient times, or more recently? When the scientists drew the family trees and went through the numbers, the answer was clear. The evidence showed that the priestly Y chromosome originated 2,100 to 3,250 years ago, sometime between the Exodus and the destruction of the First Temple—in other words, just about when the Bible says it occurred. The Cohanim genetic signature is not a recent invention, or a consequence of priests in the Middle Ages adopting the same last names. It is an indicator of an ancient tradition faithfully carried out from one generation to the next.

The DNA results offer concrete evidence that Jewish men have followed the biblical injunction to designate only the sons of Cohanim as priests over the millennia. But were their wives as faithful to their husbands as their husbands were to tradition? To find out, the scientists treated the data like a historical paternity test. The results were reassuring. Less than 0.1 percent of the Cohanim line resulted from infidelity, an amazingly low figure when compared to present-day nonpaternity rates of 5 to 10 percent, the percentage routinely found by DNA testing.

Abraham, Isaac, and Ishmael

Just as Cohanim trace their ancestry back to Aaron, Jews of every background consider themselves the descendants of Abraham. One can't help but ask, Does DNA analysis back this claim, as well? Are Jews a genetically identifiable group?

To find out, Hammer and collaborators compared the Y chromosomes of 1,371 Jewish and non-Jewish men from Africa, Asia, and Europe. Analyzing 18 different DNA markers, they showed that most Jewish populations are not significantly different from one another at the genetic level despite the fact that they have been separated for millennia. For example, Jews from Poland, Russia, Italy, Morocco, Libya, Iraq, Iran, and Yemen were more similar to one another than to their non-Jewish neighbors. Most of the genetic patterns observed in contemporary Jewish communities could be traced back to a common source population several thousand years ago.

The group that the Jews were most similar to were Middle Eastern Arabs. Non-Jewish men from Lebanon, Syria, and Palestine had virtually the same mix of Y chromosomes as did the various Jewish populations that were surveyed. Given nothing more than a DNA sample, it would be impossible to tell the difference between a typical Jew and a typical Arab.

This genetic similarity is not surprising to biblical scholars. According to Genesis, Abraham fathered his first son, Ishmael, with his wife's maid, Hagar, an Egyptian woman. Isaac, his second son, was born to Abraham's Jewish wife Sarah after God allowed her to become pregnant at an advanced age. This implies that the first Arabs and Jews had the same Y-chromosome DNA—a genetic resemblance that survives to this day.

One surprising discovery from the study was the low rate of biological intermingling between Jews and their host populations. In the case of Ashkenazic Jews, the study suggested that there was less than a 0.5 percent admixture each generation with Europeans. In other words, fewer than one out of two hundred Jewish women had sons with non-Jewish men, whether through marriage or illegitimately. It is a puzzle, then, why many Ashkenazim look so European; perhaps scientists will learn the answer from studying the DNA sequences on other chromosomes, which are more likely to carry genes involved in how the body is formed.

Another unexpected finding from the DNA analysis was the relative uniformity of Jewish Y chromosomes. If Abraham's seven Jewish sons (he had six more with his second wife, Keturah) each gave him seven grandchildren, who in turn each gave him seven great-grandchildren, and so on down the line, the current collection of chromosomes would have been more diverse. Instead, it appears that a small number of men had many sons, all with the same Y chromosome, while most men left no male descendants. The most likely explanation is that the early Jews practiced polygamy. Some men had many wives, but most had none. Polygamy is mentioned in the Bible and is still a common practice among nomadic desert people, including present-day Bedouins.

How did the Jews remain a genetically coherent group despite 3,000 years of disruptions, dispersions, exiles, expulsions, and migrations? First, when Jews did intermarry, the children usually were raised as gentiles. Jews were more often assimilated into other cultures than they assimilated others into theirs—no surprise for a people who have consistently been the target of discrimination. Second, they made relatively few converts. Judaism is, for the most part, not a proselytizing religion.

DNA analysis has revealed one fascinating exception to the above principle of assimilation. A recent study by Skorecki and colleagues uncovered a subgroup of Ashkenazic Levites who have a Y-chromosome pattern that is not seen in other priests, or indeed any major Jewish group, but is common in people around the mouth of the Volga River. A little sleuthing revealed the historical connection.

Around A.D. 600, a belligerent tribe of Mongolian people known as the Khazars conquered what is now southeastern Russia from the Caspian to the Black Sea. Sometime in the eighth century, they decided to convert from paganism to monotheism. Most of the common people became either Christian or Muslim, but the royal family and many members of the nobility opted for Judaism. They continued to rule the region for nearly five hundred years as a Jewish state. The DNA evidence shows that many of the Khazar converts declared themselves to be not only Jews but of the priestly caste. Thus the infusion of new genetic lines.

Buba's DNA

When the northern part of the land of Canaan was conquered by the Assyrians in 721 B.C., the people of the ten tribes of Israel were either eliminated, assimilated, or exiled. Many modern-day groups, including the Mormons of the United States and the Falasha of Ethiopia, claim to be descendants of these "ten lost tribes." DNA testing, however, has failed to substantiate most of these claims. The Mormons, for example, are clearly of Western European descent, whereas the Falasha show typical African DNA markers. There is one group, however, for whom DNA analysis has provided a direct link to the Jews of biblical times.

These are the Lemba, a southern African tribe who inhabit present-day South Africa and Zimbabwe. They speak Bantu and are as black as any other Africans. But there are certain cultural oddities of the Lemba: Like practicing Jews, they strictly observe a Sabbath day, they circumcise their little boys, and they do not eat any shellfish, pork, or porklike meats such as hippopotamus.

The Lemba believe they are descended from an ancient group of Jews who were led out of Israel by a prophet named Buba. According to legend, they traveled first to the city of Senna, on the Gulf of Aden in what is now Yemen, and then on to their present location in the south of Africa. They were led by the descendants of Buba, a clan that is still given deference in modern Lemba society.

Until recently, most anthropologists and biblical scholars believed that the Lemba were just one more group who derived their folklore and customs through the zeal of early Christian missionaries rather than any real genetic inheritance. But Tudor Parfitt, director of the Center for Jewish Studies at the School of Oriental and African Studies in London, thought the Lemba just might be right about their history. He has actually located what he believes to be the ruins of Senna, now a small village in a remote, previously unexplored valley of Yemen. He also believes that it is possible to account for the long journey to South Africa by the fact that a sailboat can make it from the port of Sayhut on the Gulf of Aden to the south of Africa in as little as nine days if the winds are just right.

When Parfitt heard about David Goldstein's discovery of a genetic signature for the Cohanim, he knew just what to do. He promptly collected DNA samples from the Lemba. He explained to the villagers that their DNA, extracted from swabs of cheek cells, contained "the footprints of your ancestors." Then he sent the DNA up to Oxford for analysis.

The results were remarkable. The Lemba people carry the genetic signature of the Jewish Cohanim. About 10 percent of the Lemba men had the telltale pattern of nine DNA variations that constitute the CMH, a figure as high as for lay Jews living in the Middle East and Europe. By contrast, none of the surrounding African peoples had this set of markers.

Even more incredible, fully 53 percent of the members of the Buba clan, the hereditary priestly cast of the Lemba, carried the DNA signature. That is as high a percentage as is found in the Cohanim of Jerusalem or Tel Aviv. The Y-chromosome DNA of the Buba clan is the same as the Y-chromosome DNA of Moses,

Aaron, Nadab, and Abihu. It is a remarkable testament to the power of religion that the Lemba Jews—descendants of a tiny tribe of people, born into a poor and arid land, trapped between two rich and powerful kingdoms, and dispersed against their will to the four corners of the globe—have maintained both their culture and their DNA despite thousands of years of exile from the holy land.

When Nonsense DNA Is Meaningful

The Y-chromosome DNA sequences used to trace the history of the Jews are located in a sort of insulator material that separates the functional genes from one another but doesn't code for anything itself. They're just markers, not actual genes. Although such sequences often are considered to be "nonsense" DNA, they illustrate a key point about genes and behavior: the power of culture as a genetic selective force.

Such nonsense sequences would not tell us much about most religions. If one were to look at the DNA of Catholics, for example, no specific pattern would emerge, since there are converts from every race of man.

But Judaism is different. Genetic identity is important. The reason is sociocultural—namely, the fact that Jews believe that Judaism is based on a contractual relationship between God and the people of Israel. According to Jewish belief, God did not say, "I will take care of anyone who worships me." What he promised, according to the Hebrew Bible, was that "I will take care of any *Israelite* who worships me."

There are other examples of the special relationship between God and the Jewish people. While Moses was on Mount Sinai, God instructed him that the priest's clothing should include a chestpiece with 12 different gemstones, each one set in gold and engraved with the name of one of the twelve tribes. "In this way," God said, "the Lord will be reminded of his people continually." It served as a sort of genealogical cheat sheet.

Such reminders are frequent in the Old Testament of the Bible, much of which reads like the working notes of a genealogist. For that is what page after page of "who begot whom" is all about. They are genetic records used to trace the bloodlines of the tribes of Israel—to see who was a legitimate partner to enter into the covenant with God. When I asked David Goldstein, a key player in the Y-chromosome research, what the most important implication of his work was, he had a ready answer: "It shows that the Jews are a people."

Culture and DNA

The relationship between culture and DNA is not unique to Judaism. It's also found in Hinduism, the dominant social and religious system of India. Indian men have little social mobility, because they almost never marry women of higher caste; if an Indian man marries a woman of lower caste, his status remains unchanged, according to Hindu custom. Women are more socially mobile, since their status is upgraded if they marry a man of higher caste.

To study the effects of this centuries-old system on DNA, scientists examined members of the five major caste groups using both Y-chromosome markers to study paternal inheritance and mitochondrial markers to track maternal inheritance. As expected from the arbitrariness of male social position, there was no relationship between caste and DNA markers on the Y chromosome.

By contrast, the ability of women to move from one caste to another led to a strong correlation between maternally inherited mitochondrial DNA and social rank: The closer the castes, the more similar the DNA; the farther apart the castes, the more different the DNA. Just as with the Jews, a cultural practice led to discernible differences at the genetic level.

The point of this discussion is to highlight the powerful effects of culture on genetic inheritance. Usually when we discuss natural selection we think of the physical environment. Polar bears have white coats because they were selected to survive in a snowy environment, for example, and mangroves have multiple roots to stabilize them in swampy soil. Or we think of direct effects on survival, such as the possible benefits of "God genes" on longevity discussed in Chapter Eight.

But the DNA studies of the Jews and Hindus show that sociocultural religious practices, such as allowing only the sons of priests into the priesthood or allowing only women to marry upward, can be just as powerful. This may have been the way that the intricate dance between genes that foster spirituality and religious memes started long ago.

Eleven

God Is Alive

Is the sense of Divine presence a sense of anything objectively true?

—*William James*

When I was growing up, my family received *Time* magazine every week. It was usually left out in plain view on top of the coffee table in the den. Although I must have seen hundreds of covers, I remember only one. It had no photograph or artwork, just a simple question printed in red ink on a black background: *"Is God Dead?"*

The cover story described how a group of theologians calling themselves Christian atheists had come to the conclusion that God was dead. They were led by Thomas Altizer, a professor of religion at Emory University. He predicted that as the power of science and technology waxed, faith in God would wane.

The story elicited the most reader mail ever received by *Time.* It certainly made an impression on my 13-year-old mind, perhaps because I was in the midst of an adolescent spiritual awakening—a period when I was entranced with gods, rituals, magic, and prayer. Little did I know that, some three decades later, I would be asking the same sorts of questions that fascinated Altizer and the readers of *Time*. Does science disprove religion—or does it, in fact, reveal some of the mechanism by which it works? Can science and religion peacefully coexist—or are they inevitably in conflict? Will scientific understanding replace the need for belief—or make it even greater?

Altizer's prediction was, of course, all wrong. God has not died; he's as alive as ever. Just consider what's happened in China over the past 500,000 years or so.

From Peking Then

On October 19, 1927, a tall, balding paleontologist named Birgir Bohlin strode into the office of his colleague Davidson Black, the head of the Department of Anatomy at Peking Union Medical College and an expert on dental evolution and the study of skeletons. In his hand Bohlin bore what eventually would lead to the earliest evidence for spiritual beliefs in humans. It was a tooth—a well-preserved lower left molar.

Bohlin discovered the tooth at Dragon Bone Hill, a low-lying knoll near the village of Zhoukoudian, 30 miles southwest of present-day Beijing. It is a drab area these days, enlivened only by the brilliant orange of an occasional persimmon tree. But in Bohlin's time it was bustling with activity. The hill was being excavated by a team of American, European, and Chinese scientists; many fossilized animal bones had been found there by workers in the nearby limestone quarry. The workers sold the fossils to Chinese apothecaries, where they were ground up and sold as "dragon bones."

Black was excited by the tooth. It was the first evidence of hominid life to be found after nearly a year of laborious and expensive digging and sifting. He put it into a custom-made brass canister, which he strung around his neck and took on a lecture tour of Europe and the United States. In New York he visited the Rockefeller Foundation, where he used the tooth to secure an $80,000 grant to continue the project.

It was money well invested. Over the next ten years, the stratified clay and limestone of Dragon Bone Hill yielded many more hominid remains, including jawbones, arm and leg fragments, and three complete skulls. The scientists dubbed the fossils "Peking man" and speculated that they represented the long-sought lost link between apes and man. Stone tools and signs of fire, such as charred bones and ash, were also found at the site.

Peking man was a member of the species *Homo erectus*, which evolved in Africa about 1.5 million years ago and then gradually spread into Asia. By the time he appeared in China, approximately 500,000 years ago, he was considerably smarter than his ancestors. He had almost twice as large a brain as any previous hominid and was the first to practice long-distance travel, systematic hunting, and the use of fire. Although Peking man was still a long way from modern man—it's not clear whether he even had language—he appears to have engaged in two practices that connote a mystical belief system: ritual cannibalism and skull preservation.

All of the skulls found at Zhoukoudian had been neatly cracked open at the foramen magnum, which is where the spinal cord enters the brain through the base of the skull, presumably to facilitate extracting and consuming the brains. The long bones of the arms and legs were also split open, probably to remove the marrow. Skulls were often stored separately from the remains of the bodies.

It is unlikely that Peking man practiced cannibalism out of hunger, since he was an excellent hunter, as attested to by the large number of animal bones found in and close to the cave. If Peking man didn't eat the brains for food, scientists speculate, then perhaps he did so to gain some sort of mystical power.

Anthropologists speculate that Peking man associated the brain with the life force of a person and believed that eating it would transfer that essence. This would represent an early form of sympathetic magic. If one substitutes the word "soul" for brain, the relationship to traditional spiritual beliefs becomes obvious.

The speculation is admittedly sketchy—we aren't even sure if Peking man practiced cannibalism at all, much less why. But there is support for this kind of ritual cannabalism from cultures that did leave historical records.

For the Aztecs, for example, cannibalism represented a direct route to divinity. The Aztecs believed that their gods had an insatiable hunger for fresh human flesh and would destroy the world if they didn't get their fill. Human sacrifice became a major enterprise of Aztec society; some 20,000 people were captured and sacrificed just for the opening ceremony for the great pyramid at Tenochtitlán alone. The victims were marched to the top of the pyramid, bent backward faceup over the altar, eviscerated, then rolled back down the steps. The warrior who had captured the victim was allowed to drink his blood and eat his heart. The rest of the body was taken to a temple, divided up, and eaten that night.

It has been argued that the Aztecs practiced cannibalism to increase their protein intake, but their own explanation was more religious. They believed that the victims took on the nature of a god when they were sacrificed; by eating their flesh, that power would transfer to those who ate them. It was an extreme case of "you are what you eat."

This relationship between cannibalism and mystical beliefs is not unique to the Aztecs. In India, the Binderwurs ate the dead to please the goddess Kali. In some African and North American tribes, the bodies of enemies were ingested either to absorb their virtues or to ensure that their ghosts would have nowhere to live.

It is difficult to find a society in which cannibalism is practiced without a ritual or supernatural interpretation.

Peking man's practice of storing the dead's skulls separately from their bodies may also have had significance. The notion of the skull as the home of a person's life force is still alive today in hunter-gatherer societies such as the Jivaro, a bellicose tribe inhabiting the eastern Andes of Ecuador and Peru. When a Jivaro warrior kills an enemy in battle, the victim's head is removed, boiled, dried, blackened over hot rocks, and shrunken to the size of an apple. The purpose of this ritual is to trap the spirit of the enemy inside the head so that it cannot escape and seek revenge on the warrior.

By 100,000 years ago, Peking man and his brethren had been supplanted by several species closer to modern humans. The best known is Neanderthal man, who appeared in Europe and the Near East during the last ice age. Despite the popular view of the Neanderthal as "primitive" because of his apelike body and large jaw, he actually had a considerably larger brain and more extensive tool kit than his predecessors did. More important from the religious point of view, he had begun to bury his dead.

One of the most famous Neanderthal burial sites is located in a cave near La Chapelle-aux-Saints, France. There the skeleton of a mature man with his knees drawn up to his chin was found in a shallow excavation in the grotto floor. Nearby there were a variety of flint tools, pieces of red ochre, and the bones of woolly rhinoceros, reindeer, and bison. The people who buried this man seemed to believe that he would be reborn. The body was buried in the fetal position, as if ready to emerge from the womb of the cave into a new world. The red ochre may have been used to paint the body the color of blood, like a newborn baby. It has been speculated that the tools were placed there so the reborn would have something to work with, and the bones were the remains of animal sacrifices made to provide him with something to eat. There certainly seems to be a resemblance to the more elaborate burial sites, such as the pyramids of Egypt, of cultures known to believe in an afterlife.

It's important to note that there is considerable debate about the meaning of these prehuman findings. Many anthropologists contend that they can all be explained by pure behavioralism without invoking religious or prereligious beliefs at all. The distinguished American anthropologist Marvin Harris, author of *Cannibals and Kings*, speculates that Peking man practiced cannibalism as a form of "fast food" and that Neanderthals buried the dead only to avoid the stench of decay in the cave. In truth, we will probably never know for certain whether there was religious intent behind the burial activities of Peking man and the Neanderthals. There is barely enough information to piece together what they were doing, much less what they were thinking. There is no oral history, no explicitly religious artifacts, and certainly no written record. All that is left are bones.

The evidence gets stronger with the emergence of our own species, *Homo sapiens*, which first appeared in Africa 150,000 to 200,000 years ago, in Israel and environs 100,000 years ago, and in Europe some 60,000 years later. The original humans' lifestyle was not terribly different from that of the Neanderthals. They were primarily hunters and gatherers.

Then something changed. Within an eyeblink of geological time, and for reasons that anthropologists still don't understand, man blossomed. Humans began to talk to one another. They began to paint and sculpt, to sing and dance. Soon they were farming and building permanent residences. And the evidence seems to suggest they were developing spirituality as well.

Because these early humans did not write, their convictions must again be inferred from their artifacts. But what artifacts they are! The Grotte Chauvet, a limestone cavern in the Ardèche Gorge in France, contains a series of murals that were produced some 32,000 years ago. There are dashing portraits of lions, bears, rhinos, and mammoths—animals that were not frequently hunted but may have been admired for their strength and inner power. There are engravings of pubic triangles, perhaps symbolizing fertility. There are 55 carefully arranged bear skulls, one placed on a rock altar, suggesting a cult of cave bear worship.

Most intriguing of all is a picture of a chimera with the head and torso of a bison but the legs of a human. It appears to represent an early shaman. A somewhat similar drawing from the Trois Frères cave in France depicts a dancing man wearing an animal disguise. He has a long beard, like that of a goat, on his chin, a large pair of antlers perched on his head, bear paws at the ends of his arms, and the tail of a horse. But his legs and feet are obviously human—just as one might expect to see in a man loosely draped in animal skins as he dances around a fire. This figure is commonly known as "the sorcerer" because of its resemblance to the shamans and sorcerers of modern hunter-gatherer societies. We do not know if the sorcerer danced to appease the spirits of the dead animals, to honor the hunters who killed them, or to attract more animals in the future. But whatever the precise purpose, it seems likely to have been ritual rather than practical, and most likely to have involved some form of spiritual animation.

Both the Grotte Chauvet and Trois Frères sites are located in isolated, dark caverns with no signs of habitation. Nevertheless, they were visited by people for thousands of years, perhaps on some sort of pilgrimage. These paintings were not idle drawings made to pass away a rainy day. They were made deliberately and with a great deal of effort. They were made for a spiritual purpose.

Over the next 20,000 years, signs of religious activity become increasingly rich and varied. A lakeside site in northern Germany yielded the skeletons of twelve reindeer that had been weighted down with rocks and drowned. These were likely sacrifices to the gods; it is difficult to imagine otherwise why an animal so essential for food and clothing would have been discarded. Many limestone and ivory carvings of women have been found. Their exaggerated secondary sex characteristics—gigantic breasts, round and protruding buttocks, extended abdomens, and bold labia—suggest that they represented fertility charms. A cave painting at Cogul, Spain, shows a group of women dancing around a boy, perhaps representing an early initiation rite. In a cave at Ofnet, Germany, 33 human skulls were found, all decorated with red ochre and facing west, as if looking toward heaven. In Egypt, many graves have been found near dwelling sites, some of them containing separated skulls. This predynastic practice of separating skulls and bodies was described in the *Book of the Dead*, providing one of the rare links between prehistorical and historical religious practices.

Although the prehistoric evidence of religion is often difficult to reconstruct precisely, and the details of what people believed in are open to varying interpretations, it is impossible to ignore or deny completely. By historical times, religion was an essential element of human life; indeed, it was the primary topic of many of the earliest written records. Such formalized systems of spiritual belief are at least as old as man, and perhaps even older. Even more remarkable than their antiquity, though, is their continued presence in the modern world.

To Beijing Now

In the 500,000 years between the Peking of *Homo erectus* and the Beijing of today, religion flourished in China. Confucianism, which stresses moral values and ancestor veneration, was developed in the sixth century B.C. and subsequently exported throughout East Asia. It was counterbalanced by Taoism, a metaphysical system that was developed about the same time but that emphasizes magic, mysticism, and harmony with nature. Six hundred years later, Buddhism was introduced from India, according to tradition, by the Han emperor Ming Ti, who sent emissaries to the subcontinent after dreaming of the Buddha in the form of a flying golden god. These three religions coexisted in China for 2,000 years.

When the Communists gained power in China in 1949, however, all organized religious activity was banned. Authorities burned down the old monasteries and destroyed the ancient scrolls. Churches and temples were converted to housing or government offices. Believers were separated from their families and sent to the countryside, and anyone associated with a foreign religion like Christianity was accused of spying. All overt signs of religious activity disappeared. Religion seemed to have died. Or had it?

It was not until the 1980s that the Communist Party began to relax its economic and social control. The strict limitation on family size was rescinded, and people were allowed to move within the country without an internal passport. But there were unanticipated consequences of liberalization. The government soon learned that it takes more than a few decades of repression to erase millennia of spiritual tradition. Their unexpected new enemy was a homegrown religious movement: Falun Gong.

Falun Gong is a New Age conglomeration of Taoist and Buddhist beliefs combined with *qigong,* an ancient Chinese spiritual system that focuses on the body's vital energy, or *qi.* It combines graceful, slow-motion exercises with meditation, healing practices, and a strict behavioral code. Adherents believe that practicing Falun Gong creates an orb of energy in the belly that improves their health and well-being and ultimately allows them to evolve into a different form of matter. There are claims that it can be used to heal fatal diseases and move objects by telekinesis.

Falun Gong has no churches or temples. Believers gather in public parks or plazas to exercise and read together. Although they come from all walks of life, the most numerous devotees seem to be middle-aged and elderly women. They often perform their exercises early in the morning, dressed in jogging suits and sneakers.

The founder of Falun Gong is Li Hongzhi, a former clerk at a grain company. Li claims that he can insert a "wheel of law" into his disciples' bellies by telekinesis. He also believes that the world is gradually being infiltrated by multicolored aliens, and that there are separate paradises for each race. He claims to be able to fly, and to make himself invisible.

Li began lecturing and promoting Falun Gong in the late 1980s. At first he was allowed to proselytize freely. But as the movement picked up steam, gaining as many as 10 million adherents by the mid-1990s, the government became suspicious. In 1997, they banned Li's books and curtailed meetings of his followers. This led to a series of challenges and counterchallenges that culminated on April 15, 1999, when an estimated 10,000 Falun Gong followers surrounded the Communist leaders' compound in Beijing in silent protest. It was the largest antigovernment demonstration since the Tiananmen Square pro-democracy movement ten years previously.

The government was not amused. They banned Falun Gong as an unauthorized cult. Thousands of followers were arrested, jailed, and in some cases tortured. Leaders were sentenced to long prison terms, and Li Hongzhi was declared an enemy of the state. Propaganda was issued, including a videotape titled *Falun Gong—Cult of Evil.*

The reaction was all too predictable. Li Hongzhi became a worldwide hero, and his following grew. Meanwhile, the Communist leaders became an international laughingstock for their brutal suppression of "a bunch of old ladies doing calisthenics." Falun Gong was glorified, not eradicated.

The rise of Falun Gong in China is just one of countless examples of the strength and durability of religion. Political prohibitions, such as those imposed throughout the Communist world during much of the past century, have never been capable of diminishing religion's appeal. Neither technological advances nor economic progress has diminished the beliefs of the faithful. Today, although the United States is one of the most technically sophisticated and wealthiest countries in the world, it is also one of the most religious.

The news of God's death announced by Thomas Altizer and the Christian atheists in 1965 was, to paraphrase Mark Twain, premature. God is alive and well.

The Battle Between Science and Religion

Religion without science is blind; science without religion is lame.

—*Albert Einstein*

When I told my former boss at the National Institutes of Health that I was writing this book, her suggestion was "Wait until you've retired." Her attitude is not unusual; most scientists regard interest in spirituality and religion as a sign of bias or nonobjectivity, if not downright senility. Theologians, for their part, often see science as irrelevant, incomprehensible, or even destructive. There doesn't seem to be much common ground.

That's a pity, for as Einstein noted, there are many ways in which science and religion complement each other. Recently, an unlikely peace broker has appeared on the scene: John Marks Templeton, a Tennessee-born financier, has used a good chunk of his fortune to finance the Templeton Foundation. The foundation supports research projects, academic courses, writing, and conferences on the benefits of cooperation between science and religion. The most visible activity of the foundation is the annual Templeton Prize for Progress in Religion, which at more than one million dollars is the world's most generous award.

Templeton's efforts have led to renewed interest and academic inquiry in science and religion; many of the research studies cited in this book, in fact, were supported by his foundation. By and large, however, the intersection between science and religion still resembles a battlefield. There are those who claim that science disproves religion, others who believe that religion trumps science, and those who say there is no conflict—even as they duck the incoming shells.

The scientific method, of course, is based on observation, experimentation, and replication—methods that exclude most religious phenomena from consideration. There is no way to objectively test, for example, whether the consciousness continues after death, or whether God listens to our prayers. There is no MRI or CAT scan for the human soul. It is therefore not surprising that many scientists regard religion with suspicion.

Richard Dawkins, the evolutionary biologist described earlier, is an extreme example, as his words will show:

> There is no reason for believing that any sort of gods exist and quite good reason for believing that they do not exist and never have. It has all been a gigantic waste of time and a waste of life. It would be a joke of cosmic proportions if it weren't so tragic.

Dawkins is at his most effective when he takes on the "argument by design" for God. This is the religious belief that the universe and the creatures in it are too perfect to have arisen by accident, leading to the conclusion that there must have been an engineer—God—who planned them out in advance. Using eloquent logic and compelling examples, Dawkins explains why this theory is superfluous and makes a convincing case that evolution is capable of explaining all life-forms, including our own. But he then takes the argument one step too far by saying that because evolution by itself *can* explain life, therefore it does.

This is an elementary flaw of logic. It is a perfect example of the hoary misconception that if A can cause B, then every B must have been caused by A—and A alone. Most scientists feel evolution is sufficient to explain life without a designer, but from a purely scientific viewpoint, it by no means proves that there is no designer. Ironically, then, it appears that Dawkins does have a religion—science—which he follows based on his own faith rather than logic.

Religion is based on exactly the opposite principle as science. It depends on faith—belief even in the absence of tangible evidence. Many of the specific tenets of the world's major religions, all of which were formed more than a millennium ago, seemingly run directly counter to modern science. This raises a serious dilemma for believers: Which is true, science or religion?

Historically, each time religious contoversy has touched science, it has been theology that's been burned. In the sixteenth century, the church taught that the sun circled the earth; astronomy proved it to be the other way around. In the seventeenth century, infectious diseases were seen as punishments by God; microbiology taught us their true origin. In the ninteenth century, spontaneous generation was accepted dogma; it took Pasteur to show that life comes only from life. In the United States today, the most visible such clash has occurred over the issue of evolution and creationism. A literal reading of the Bible teaches that the universe, the earth, and all its flora and fauna were formed in six days just a few thousand years ago, and that man was created in the image of God. Scientific evidence, on the other hand, shows that the universe was created billions of years ago, that life gradually arose by the process of evolution, and that humans have evolved from other species.

Religious fundamentalists' insistence that the biblical version of creationism is literally true has led to much frantic hand-waving to explain away the scientific findings. To explain why humans have the same genetic code as every other life-form, for instance, they suggest that God deliberately made our biochemistry the same. The presence of fossils, oil, diamonds, and countless other signs of the earth's great age is likewise discounted as deliberately planted evidence intended to mislead the unfaithful. One creation scientist has even tackled the problem of why it is possible to detect, through powerful telescopes, light from distant galaxies that has been traveling through the solar system for billions of years. His explanation is that God put the earth at the center of a giant black hole that accelerates the speed of light, thus compressing billions of years into a few thousand.

Of course, the majority of those who believe in the Bible do not interpret it literally. Most people appreciate science for its elegance and ability to understand how the world works, without assuming that such understanding in any way undermines belief in a higher being, whether we call that being God, Buddha, Muhammad, or whatever.

But I also know, from the reactions I have received to my writing about religion and science, that it is an emotional topic. The first time I wrote about the idea of "the God gene," in a diary for the online magazine *Slate,* the flood of reader responses was split into two camps. Half of them criticized me for being deterministic and reductionist with no spiritual sensitivity. The other half criticized me for not realizing that there's no such thing as a God gene—just a gene for stupidity, which I, according to these correspondents, clearly possessed.

What will it take to achieve a reconciliation between science and religion? There have been many thoughtful analyses of the differences, similarities, and overlaps between the two areas, contributions from Ian Barbour, John Polkinghorne, Émile Durkheim, Bertrand Russell, and Richard Feynman among them. I close with three ideas that reflect my perspective as a biologist and geneticist.

Softwired for God

All religions are true in their own way.

—*Émile Durkheim*

First, it is essential to realize that there is nothing intrinsically theistic or atheistic about postulating a specific genetic and biochemical mechanism for spirituality. If God does exist, he would need a way for us to recognize his presence. Indeed, many religious believers have interpreted the brain-scan experiments of Persinger, Ramachandran, and Newberg as supporting the existence of a deity; why else, they ask, would we have a "God module" prewired in the brain? If, on the other hand, there is no God, then all religious and mystical experiences may represent no more than the random firing of poorly programmed neurons in the brain. As I cautioned at the beginning of this book, science can tell us whether there are God genes but not whether there is a God. Spiritual experiences, like all experiences, must at some level be interpreted by our biologically constructed brains.

The second point to be emphasized is that just because spirituality is partly genetic doesn't mean it is hardwired. Our genes are more like a family recipe handed down by word of mouth than a precise set of instructions that must be followed in exact detail. The final product depends a lot on how you interpret and execute the formula. Everyone knows that cooking takes practice. Give eggs, butter, and herbs to an amateur and you get scrambled eggs. Put the same ingredients in the hands of an experienced chef and voilà!—a scrumptious *omelette aux fines herbes*.

It's the same with spirituality: spiritual enlightenment takes practice. Recently C. J. Hamerl, a psychology student at the University of London, studied the relationship between contemplative practice and self-transcendence. He gave a questionnaire that measures the three character traits of Cloninger's biosocial model to 159 participants recruited from Buddhist meditation centers in the London area. There was a clear relationship between meditation experience and self-transcendence. Control subjects who were interested in meditation but had not yet started had the lowest scores, followed by beginning meditators, and finally by experienced meditators with more than two years of practice. The correlation between how much the subjects meditated and their sense of self-transcendence was highly significant, and greater than that for any of the other scales measured.

This experiment implies that it's possible to strengthen one's sense of spirituality by practicing it—an idea that has been stressed by every great spiritual leader from Buddha to Gandhi, from Muhammad to Martin Luther King Jr., from Jesus to Bishop Ruis. We have no say over the exact constellation of genes that we inherit; whether we have a C or an A at the critical position in VMAT2, the genes of a Tenkai or of a Hannibal Lecter, is purely a matter of chance. What we do with our spiritual genes, however, is very much up to us.

Seeking Consilience

My final point concerns a distinction that is all too often overlooked in the debate between believers and scientists: the fundamental difference between spirituality and religion.

Claiming to be spiritual even though not religious has admitedly become something of a cliché. (My favorite example is Monica Lewinsky's response to an interviewer who asked if she thought her relationship with Bill Clinton was sinful. "I'm not very religious," she replied. "I'm more spiritual.") But there are real differences. Spirituality is based in consciousness, religion in cognition. Spirituality is universal, whereas cultures have their own forms of religion. I would argue that the most important contrast is that spirituality is genetic, while religion is based on culture, traditions, beliefs, and ideas. It is, in other words, mimetic. This is one reason why spirituality and religion have such differing impacts on individual lives and society.

The idea that spirituality is inherited is not new. A Tungus shaman cited in Joseph Campbell's *The Masks of God* says, "A person cannot become a shaman if there have been no shamans in his sib." As we've seen from twin studies, self-transcendence, a quantifiable measure of spirituality, is partially inherited. At least one of the God genes, VMAT2, appears to code for a protein that controls the ebb and flow of monoamines, brain chemicals that play a key role in emotions and consciousness.

This link makes sense. Spirituality is very much about the way we perceive the world and our role in it—processes that are mediated by consciousness. Altering a person's consciousness can help that person to realize that he or she is not necessarily at the center of the universe, that things are not always as they seem. It makes little difference whether the alteration was caused by a DNA sequence variation at the VMAT2 gene, a drug like psilocybin, or a mystical experience achieved after years of meditation. It's all about seeing the world through new eyes.

The fact that spirituality has a genetic component implies that it evolved for a purpose. No matter how selfish a gene is, it still needs a human being as a carrier to perpetuate itself. There is now reasonable evidence that spirituality is in fact beneficial to our physical as well as mental health. Faith may not only make people feel better, it may actually make them better people.

Religion is based on memes, traditions of culture that are passed on not by DNA but by learning, instruction, and imitation. As a result, memes may or may not be beneficial to the people who hold them. What determines their survival is how well they are transmitted from one person to another. The deep and abiding appeal of religious memes is based not on logic or even benefit to our species, but on their ability to evoke the altered states of consciousness and spirituality that are so important in spirtuality.

It is for this reason that religions survive, regardless of whether they advance or hinder human life and society. Many faith-based organizations provide essential social services, and churches and temples are vital places to get in touch with spirituality and to seek guidance in life and solace from life's calamities. But established religions have had their negative sides, too.

Historically, more than a few religions have promoted intolerance and have fought against human independence and freedom.

Historically, religion made a flat earth the center of the universe and fought against many scientific advances. Change has been slow. Galileo's excommunication was annulled only a few years ago, and many churches and religions continue to oppose the teaching of evolution.

Historically, religion has at times helped to usher in wars, crusades, and jihads. Today, we have September 11—the result of beliefs based on an extreme interpretation of Muslim texts and teachings.

Are we condemmed to endlessly repeat the sad mistakes of the past? I believe that understanding the difference between spirituality and religion gives us another tool in offering hope for the future. For while our spirituality may be engraved to a degree in our DNA, we can change, reinterpret, and reconsider the memes written on the scrolls of our religious history.

Our genes can predispose us to believe. But they don't tell us what to believe in. Our faith is part of our cultural heritage, and some of the beliefs in any religion evolve over time. Some of religion's least desirable memes, such as the condemnation of pagans, of non-believers, of outsiders, can be difficult to erase or reinvent. But they can be altered—and in the case of religious memes that prove themselves to be destrucive to peace, understanding, and compassion, they *must* be.

Understanding the distinction between spirituality and religion may yet bring an armistice in the win/lose debate between science and faith. But it is important to distinguish between the beliefs and the act of believing—which is one of the greatest gifts of being human.

Sources and Further Reading

Chapter I: Spiritual Instinct

Armstrong, K. *A History of God*. New York: Ballantine Books, 1994.

Doninger, W. (Ed.). *Encyclopedia of World Religions.* Springfield, MA: Merrian Webster, 1999.

Eastman, R. *The Ways of Religion.* New York: Oxford University Press, 1999.

Eliade, M. *A History of Religious Ideas: Volume 1.* Chicago: The University of Chicago Press, 1988.

Eliade, M. *A History of Religious Ideas: Volume 2.* Chicago: The University of Chicago Press, 1988.

Eliade, M. *A History of Religious Ideas: Volume 3.* Chicago: The University of Chicago Press, 1988.

Gallagher, W. *Working on God.* New York: Random House, Inc., 1999.

Gallup Jr., G., J. Castelli. *The People's Religion.* New York: Macmillan Publishing Company, 1989.

Humpreys, C. *Buddhism: An Introduction and Guide.* New York: Penguin Books, 1990.

James, W. *The Varieties of Religious Experience.* New York: Touchstone, 1997.

Kosmin, B.A., S.P. Lachman. *One Nation Under God.* New York: Harmony Books, 1993.

Sato, K. *The Zen Life.* New York: John Weatherhill, Inc., 1987.

Smith, H. *The World's Religions.* San Francisco: HarperCollins, 1991.

de Waal Malefijt, A. *Religion and Culture.* Prospect Heights, IL: Waveland Press, 1989.

Wulff, D.M. *Psychology of Religion: Classic and Contemporary.* New York: John Wiley & Sons, 1997.

Chapter 2: Self-Transcendence

Cloninger, C.R. "A Unified Biosocial Theory of Personality and Its Role in the Development of Anxiety States." *Psych Dev* 3, 167–226, 1986.

Cloninger, C.R. "A Systematic Method for Clinical Description and Classification of Personality Variants." *Arch Gen Psych* 44, 573–88, 1987.

Cloninger, C.R. "Temperament and Personality." *Current Opinion in Neurobiology* 4, 266–73, 1994.

Cloninger, C.R., T.R. Przybeck, D.M. Svrakic. "The Tridimensional Personality Questionnaire: U.S. Normative Data." *Psych Rep* 69, 1047–57, 1991.

Cloninger, C.R., D.M. Svrakic, T.R. Przybeck. "A Psychobiological Model of Temperament and Character." *Arch Gen Psych* 50, 975–90, 1993.

Cloninger, C.R., N.M. Svrakic, D.M. Svrakic. "Role of Personality Self-Organization in Development of Mental Order and Disorder." *Devel Psychopath* 9, 881–906, 1997.

Cloninger, C.R., C. Bayon, D.M. Sravkic. "Measurement of Temperament and Character in Mood Disorders: A Model of Fundamental States as Personality Types." *Affect Dis* 51, 21–32, 1998.

Earlywine, M., P.R. Finn, J.B. Peterson, R.O. Pihl. "Factor Structure and Correlates of the Tridimensional Personality Questionnaire." *Stud Alc* 53, 233–38, 1992.

Maslow, A.H. "The 'Core-Religious' or 'Transcendent' Experience." *The Highest State of Consciousness.* Garden City, NY: Anchor, 352–64, 1972.

Maslow, A.H. *The Farther Reaches of Human Nature.* New York: Penguin Arkana, 1993.

Maslow, A.H. *Toward a Psychology of Being.* New York: John Wiley & Sons, 1999.

Mendlowicz, M.V., J-L. Girardin, J.C. Gillin, H.S. Akiskal, L.M. Furlanetto, M.H. Rapaport, J.R. Kelsoe. "Sociodemographic Predictors of Temperament and Character." *Psych Res* 34, 221–26, 2000.

Nagoshi, C.T., D. Walter, C. Muntaner, C.A. Haerrtzen. "Validation of the Tridimensional Personality Questionnaire in a Sample of Male Drug Users." *Person Ind Diff* 13, 401–9, 1992.

Spence, J.D. *The Memory Palace of Matteo Ricci*. New York: Penguin, 1986.

Svrakic, D.M., C. Whitehead, T.R. Przybeck, C.R. Cloninger. "Differential Diagnosis of Personality Disorders by the Seven-Factor Model of Temperament and Character." *Arch Gen Psych* 50, 991–98, 1993.

Waller, N.G., S.O. Lilienfeld, A. Tellegen, D.T. Lykken. "The Tridimensional Personality Questionnaire: Structural Validity and Comparison with the Multidimensional Personality Questionnaire." *Multivar Behav Res* 26, 1–23, 1991.

Chapter 3: An Inherited Predisposition

Bouchard Jr., T.J., D.T. Lykken, M. McGue, N.L. Segal, A. Tellegen. "Sources of Human Psychological Differences: The Minnesota Twins Reared Apart." *Science* 250, 223–38, 1990.

Bouchard Jr., T.J., M. McGue, D. Lykken, A. Tellegen. "Intrinsic and Extrinsic Religiousness: Genetic and Environmental Influences and Personality Correlates." *Twin Res* 2, 88–98, 1999.

Bouchard Jr., T.J., and M. McGue. "Genetic and Environmental Influences on Human Psychological Differences." *J Neurobiol* 54, 4–45, 2003.

D'Onofrio, B.M., L.J. Eaves, L. Murrelle, H.H. Maes, B. Spilka. "Understanding Biological and Social Influences on Religious Affiliation, Attitudes, and Behaviors: A Behavior Genetic Perspective." *J Personal* 67, 953–83, 1999.

Eaves, L.J., H.J. Eysenck, N.G. Martin. *Genes, Culture, and Personality: An Empirical Approach*. New York: Academic Press, 1989.

Eaves, L.J., B.M. D'Onofrio, R. Russell. "Transmission of Religion and Attitudes." *Twin Res* 2, 59–61, 1999.

Eaves, L., A. Heath, N. Martin, H. Maes, M. Neale, K. Kendler, K. Kirk, and L. Corey. "Comparing the Biological and Cultural Inheritance of Personality and Social Attitudes in the Virginia 20,000 Study of Twins and Their Relatives." *Twin Res* 2, 62–80, 1999.

Kirk, K.M., H.H. Maes, M.C. Neale, A.C. Heath, N.G. Martin, L.J. Eaves. "Frequency of Church Attendance in Australia and the United States: Models of Family Resemblance." *Twin Res* 2, 99–107, 1999.

Kirk, K.M., L.J. Eaves, and N.G. Martin. "Self-Transcendence as a Measure of Spirituality in a Sample of Older Australian Twins." *Twin Res* 2, 81–87, 1999.

Loehlin, J.C. *Genes and Environment in Personality Development.* Newbury Park, CA: Sage, 1992.

Plomin, R., J.C. De Fries, G.E. McClearn. *Behavioral Genetics: A Primer* (2nd edition). New York: W. H. Freeman, 1990.

Rowe, D.C. *The Limits of Family Influence.* New York: The Guilford Press, 1994.

Waller, N.G., B.A. Kojetin, T.J. Bouchard, Jr., D.T. Lykken, A. Tellegen. "Genetic and Environmental Influences on Religious Interests, Attitudes, and Values: A Study of Twins Reared Apart and Together." *Am Psychol Soc* 1, 138–42, 1990.

Wright, L. *Twins and What They Tell Us About Who We Are.* New York: John Wiley & Sons, Inc., 1997.

Chapter 4: The God Gene

Benjamin, J., L. Li, C. Patterson, B.D. Greenberg, D.L. Murphy and D.H. Hamer. "Population and Familial Association Between the D4 Dopamine Receptor Gene and Measures of Novelty Seeking." *Nat Genet* 12, 81–84, 1996.

Comings, D.E., N. Gonzales, G. Saucier, J.P. Johnson and J.P. MacMurray. "The DRD4 Gene and the Spiritual Transcendence Scale of the Character Temperament Index." *Psychiatr Genet* 10, 185–89, 2000.

Comings, D. E., R. Gade-Andavolu, N. Gonzalez, S. Wu, D. Muhleman, H. Blake, M.B. Mann, G. Dietz, G. Saucier, J.P. MacMurray. "A Multivariate Analysis of 59 Candidate Genes in Personality Traits: The Temperament and Character Inventory." *Clin Genet* 58, 375–85, 2000.

Gonzalez, A.M., D. Walther, A. Pazos, G.R. Uhl. "Synaptic Vesicular Monoamine Transporter Expression: Distribution and Pharmacologic Profile." *Brain Res Mol Brain Res* 22, 219–26, 1994.

Greenberg, B. D., Q. Li, F.R. Lucas, S. Hu, L.A. Sirota, J. Benjamin, K.P. Lesch, D. Hamer, D.L. Murphy. "Association Between the Serotonin Transporter Promoter Polymorphism and Personality Traits in a Primarily Female Population Sample." *Am J Med Genet* 96, 202–16, 2000.

Hu, S., C.L. Brody, C. Fisher, L. Gunzerath, M.L. Nelson, S.Z. Sabol, L.A. Sirota, S.E. Marcus, B.D. Greenberg, D.L. Murphy, D.H. Hamer. "Interaction Between the Serotonin Transporter Gene and Neuroticism in Cigarette Smoking Behavior." *Mol Psych* 5,181–88, 2000.

Myakishev, M.V., Y. Khripin, S. Hu, D.H. Hamer. "High-Throughput

SNP Genotyping by Allele-Specific PCR with Universal Energy-Transfer-Labeled Primers." *Genome Res* 11, 163–69, 2001.

Reif, A., K.P. Lesch. "Toward a Molecular Architecture of Personality." *Behav Brain Res* 139, 1–20, 2003.

Sabol, S.Z., M.L. Nelson, C. Fisher, L. Gunzerath, C.L. Brody, S. Hu, L.A. Sirota, S.E. Marcus, B.D. Greenberg, F.R.T. Lucas, J. Benjamin, D.L. Murphy, D.H. Hamer. "A Genetic Association for Cigarette Smoking Behavior." *Health Psychol* 18, 7–13, 1999.

Surratt, C.K., A.M. Persico, X.D. Yang, S.R. Edgar, G.S. Bird, A.L. Hawkins, C.A. Griffin, X. Li, E.W. Jabs, G.R. Uhl. "A Human Synaptic Vesicle Monoamine Transporter cDNA Predicts Posttranslational Modifications, Reveals Chromosome 10 Gene Localization and Identifies TaqI RFLPs." *FEBS Lett* 318, 325–30, 1993.

Uhl, G.R., S. Li, N. Takahashi, K. Itokawa, Z. Lin, M. Hazama, I. Sora. "The VMAT2 Gene in Mice and Humans: Amphetamine Responses, Locomotion, Cardiac Arrhythmias, Aging, and Vulnerability to Dopaminergic Toxins." *Faseb J* 14, 2459–65, 2000.

Vink, J.M., D.I. Boomsma. "Gene Finding Strategies." *Biol Psychol* 61(1–2):53–71, 2002.

Chapter 5: Monoamines and Mysticism

Doblin, R. "Pahnke's Good Friday Experiment: A Long-Term Follow-up and Methodological Critique." *J Transpers Psychol.* 23, 1991.

Harner, M. *Hallucinogens and Shamanism.* New York: Oxford University Press. 1973.

Malmgren, J. "Tune In, Turn On, Get Well." *St. Petersburg Times,* Nov. 27, 1994.

Pahnke, W.N. "Drugs and Mysticism." *Int J Psychol* 8, 295–313, 1966.

Pahnke, W.N., W. Richards. "Implications of LSD and Experimental Mysticism." *J Transpers Psych* 1, 69–102, 1969.

Vollenweider, F.X. "Advances and Pathophysiological Models of Hallucinogenic Drug Action in Humans: A Preamble to Schizophrenia Research." *Pharmapsychiatry* 31, 92–103,1998.

Vollenweider, F.X., M.F.I. Vollenweider-Scherpenhuyzen, A. Bäbler, H. Vogel, D. Hell. "Psilocybin Induces Schizophrenia-like Psychosis in Humans Via a Serotonin-2 Agonist Action." *Neuroreport* 9, 3897–3902, 1998.

Vollenweider, F.X., P. Vontobel, D. Hell, K.L. Leenders. "5-HT Modulation of Dopamine Release in Basal Ganglia in Psilocybin-Induced Psychosis in Man—A PET Study with [11C]raclopride." *Neuropsychopharmacology* 20, 424–33,1999.

Chapter 6: The Way Things Seem

Accili, D., C.S. Fishburn, J. Drago, H.Steiner, J.E Lachowicz, B.H. Park, E.B. Gauda, E.J. Lee, M.H. Cool, D.R. Sibley, C.R. Gerfen, H. Westphal, S. Fuch. "A Targeted Mutation of the D3 Dopamine Receptor Gene Is Associated with Hyperactivity in Mice." *Proc Natl Acad Sci USA* 93, 1945–49, 1996.

Appel, J.B., P.M. Callahan. "Involvement of 5-HT Receptor Subtypes in the Discriminative Stimulus Properties of Mescaline." *Eur J Pharmacol* 159, 41–46, 1989.

Bozzi, Y., D. Vallone, E. Borelli. "Neuroprotective Role of Dopamine Against Hippocampal Cell Death." *J Neurosci* 20, 8643–49, 2000.

Cooper, J.R., F.E. Bloom, R.H. Roth. *The Biochemical Basis of Neuropharmacology.* New York: Oxford University Press, 6th edition, 1991.

Duluwa, S.C., D.K. Grandy, M.J. Low, M.P. Paulus, M.A. Geyer. "Dopamine D4 Receptor-Knock-Out Mice Exhibit Reduced Exploration of Novel Stimuli." *J Neurosci* 19, 9550–56, 1999.

Edelman, G.M. *Bright Air, Brilliant Fire.* New York: Basic Books, 1992.

El-Ghundi, M., S.R. George, J. Drago, P.J. Fletcher, T. Fan, T. Nguyen, C. Liu, D.R. Sibley, H. Westphal, B.F. O'Dowd. "Distribution of Dopamine D1 Receptor Gene Expression Attenuates Alcohol-Seeking Behavior." *Eur J Pharmacol* 353, 149–58, 1998.

Fabre, V., A. Beaufour, A. Evrard, A. Rioux, N. Hanoun, K.P. Lesch, D.L. Murphy, L. Lanfumey, M. Hamon, M.P. Martres. "Altered Expression and Functions of Serotonin 5-HT/1A and 5-HT/1B Receptors in Knock-out Mice Lacking the 5-HT Transporter." *Eur J Neurosci* 12, 2299–2310, 2000.

Fauchey, V., M. Jaber, M.G. Caron, B. Bloch, C. Le Moine. "Differential Regulation of the Dopaminine D1, D2, and D3 Receptor Gene Expression and Changes in the Phenotype of the Striatal Neurons in Mice Lacking the Dopamine Transfer." *Eur J Neurosci* 12, 19–26, 2000.

Fei, X., R.R. Gainetdinov, W.C. Westel, S.R. Jones, L.M. Bohn, G.W. Miller, Y. Wang, M.G. Caron. "Mice Lacking the Norepinephrine

Transporter Are Supersensitive to Psychostimulants." *Nature Neurosci* 3, 465–71, 2000.

Fon, E.A., E.N. Pothos, B. Sun, N. Killeen, D. Sulzer, R.H. Edwards. "Vesicular Transport Regulates Monoamine Storage and Release but Is Not Essential for Amphetamine Action." *Neuron* 19, 1271–83, 1997.

Forman, R.K.C. *Mysticism, Mind, Consciousness.* Albany: State University of New York Press, 1999.

Gainetdinov, R.R., S.R. Jones, M.G. Caron. "Functional Hyperdopaminergia in Dopamine Transporter Knock-Out Mice." *Bio Psychiatry* 46, 303–11, 1999.

Giros, B., M. Jaber, S.R. Jones, R.M. Wrightman, M.G. Caron. "Hyperlocomotion and Indifference to Cocaine and Amphetamine in Mice Lacking the Dopamine Transporter." *Nature* 379, 606–12, 1996.

Heyser, C.J., A.A. Fienberg, P. Greengard, L.H. Gold. "DARPP-32 Knock-Out Mice Exhibit Impaired Reversal Learning in a Discriminated Operant Task." *Brain Research* 867, 122–30, 2000.

Joseph, R. *The Transmitter to God: The Limbic System, the Soul, and Spirituality.* San Jose, CA: University Press California, 2001.

Jones, S.R., R.R. Gainetdinov, X.T. Hu, D.C. Cooper, R.M. Wrightman, F.J. White, M.G. Caron. "Loss of Autoreceptor Functions in Mice Lacking Dopamine Transporter." *Nat Neurosci* 2, 649–55, 1999.

Kaori, I., I. Sora, C.W. Schindler, M. Itokawa, N. Takahashi, G.R. Uhl. "Heterozygous VMAT2 Knockout Mice Display Prolonged QT Intervals: Possible Contributions to Sudden Death." *Mol Brain Res* 71, 354–57, 1999.

Mani, S.K., A.A. Fienberg, J.P. O'Callaghan, G.L. Snyder, P.B. Allen, P.K. Dash, A.N. Moore, A.J. Mitchell, J. Bibb, P. Greengard, B.W. O'Malley. "Requirement for DARPP-32 in Progesterone-Facilitated Sexual Receptivity in Female Rats and Mice." *Science* 287, 1053–55, 2000.

Rocha, A.B., F. Fumagalli, R.R. Gainetdinov, S.R. Jones, R. Ator, B. Giros, G.W. Miller, M.G. Caron. "Cocaine Self-Administration in Dopamine-Transporter Knockout Mice." *Nature Neurosci* 1, 132–37, 1998.

Shih, J.C., K. Chen, M.J. Ridd. "Monoamine Oxidase." *Neuroscience* 22, 197–217, 1999.

Srinivasan, R., D.P. Russell, G.M. Edelman, G.Tononi. "Increased Synchronization of Neuromagnetic Responses During Conscious Perception." *J Neurosci* 19, 5435–48, 1999.

Takahashi, N., L.L. Miner, I. Sora, H. Ujike, R.S. Revay, V. Kostic, V. Jackson-Lewis, S. Przedborski, G.R. Uhl. "VMAT2 Knockout Mice: Heterozygotes Display Reduced Amphetamine-Conditioned Reward, Enhanced Amphetamine Locomotion, and Enhanced MPTP Toxicity." *Neurobiology* 94, 9938–43, 1997.

Tononi, G., R. Srinivasan, D.P. Russell, G.M. Edelman. "Investigating Neural Correlates of Conscious Perception by Frequency-tagged Neuromagnetic Responses." *Neurobiology* 95, 3198–3203, 1998.

Wang, Y., R.R. Gainetdinov, F. Fumagalli, F. Xu, S.R. Jones, C.B. Block, G.W. Miller, R.M. Wightman, M.G. Caron. "Knock-out of the Vesicular Monoamine Transporter 2 Gene Results in Neonatal Death and Supersensitivity to Cocaine and Amphetamine." *Neuron* 19, 1285–96, 1997.

Xiaoxi, Z., R.S. Oosting, S.R. Jones, R.R. Gainetdinov, G.W. Miller, M.G. Caron, R. Hen. "Hyperactivity and Impaired Response Habituation in Hyperdopaminergic Mice." *PNAS* 98, 1982–87, 2001.

Chapter 7: How the Brain Sees God

Austin, J.H. *Zen and the Brain.* Cambridge, MA: The MIT Press, 1999.

Alper, M. *The "God" Part of the Brain: A Scientific Interpretation of Human Spirituality and God.* Brooklyn, New York: Rouge Press, 2000.

Bear, D., K. Levin, D. Blumer, et al. "Interrictal Behavior in Hospitalized Temporal Lobe Epilectics: Relationships to Idiopathic Psychiatric Syndromes." *J Neurol, Neurosurgery and Psychiatry* 45, 481–88, 1982.

Ciringnotta, F., C. Todesco, E. Lugaresi. "Temporal Lobe Epilepsy with Ecstatic Seizures (So-called Dostoyevsky Epilepsy)." *Eplilepsia* 21, 705–10, 1980.

d'Aquili, E., A.B. Newberg. *The Mystical Mind: Probing the Biology of Religious Experience.* Minneapolis: Fortress Press, 1999.

Dewhurst, K., A. Beard. "Sudden Religious Conversion in Temporal Lobe Epilepsy." *Br J Psych* 117, 497–507, 1970.

Dodrill, C., L. Batzel. "Interrictal Behavioral Features of Patients with Epilepsy." *Epilepsia* 27, 564–76, 1986.

Frost, J., H. Mayberg, R. Fisher, et al. "Mu-opiate Receptors Measured by the Positron Emission Tomography Are Increased in Temporal Lobe Epilepsy." *Ann Neurol* 23, 231–37, 1988.

Hunt, H. "A Cognitive Psychology of Mystical and Altered-State Experience." *Perceptual and Motor Skills* 58, 467–513, 1984.

Maupin, E. "Individual Differences in Response to a Zen Meditation Exercise." *J Consul Clin Psychol* 29, 139–45, 1965.

Mesulam, M-M. "Dissociative States with Abnormal Temporal Lobe EEG: Multiple Personality and the Illusion of Possession." *Arch Neurol* 38, 176–81, 1981.

Newberg, A., E. d'Aquili, V. Rause. *Why God Won't Go Away.* New York: Ballantine Books, 2001.

Persinger, M.A. "Striking EEG Profiles from Single Epsiodes of Glossolalia and Transcendental Meditation." *Percep Motor Skills* 58, 127–33, 1984.

Persinger, M.A. *Neuropsychological Bases of God Beliefs.* New York: Praeger Publishers, 1987.

Ramachandran, V.S., S. Blakeslee. *Phantoms in the Brain.* New York: William Morrow and Co., 1998.

Rodin, E., S. Schmaltz. "The Bear-Fedio Personality Inventory and Temporal Lobe Epilepsy." *Neurology* 34, 591–96, 1984.

Tucker, D., R. Novelly, P. Walker. "Hyperreligiosity in Temporal Lobe Epilepsy: Redefining the Relationship." *J Nerv Ment Dis* 175, 181–84, 1987.

Chapter 8: Evolving Faith

Benor, D.J. "Survey of Spiritual Healing Research." *Comp Med Res* 4, 9–33, 1990.

Benson, H. *The Mind/Body Effect.* New York: Simon & Schuster, 1979.

Benson, H. *Beyond the Relaxation Response.* New York: Times Books, 1984.

Benson, H. *Timeless Healing.* New York: Scribner, 1996.

Boyd, R., P.J. Richerson. "Group Selection Among Alternative Evolutionarily Strable Strategies." *J Theoret Biol* 145, 331–42, 1990.

Brown, W.A. "Placebo as a Treatment for Depression." *Neuropsychopharmacology* 10, 265–69, 1988.

Chagon, N.A. *Yanomanö* (5th Edition). Orlando, FL: Harcourt Brace & Company, 1997.

Dawkins, R. *The Selfish Gene.* New York: Oxford University Press, 1976.

Dossey, L. *Healing Words: The Power of Prayer and the Practice of Medicine.* San Francisco: HarperCollins, 1993.

Eddy, M.B.G. *Science and Health with Key to the Scriptures.* Boston: The First Church of Christ Scientist, 1994.

Eliade, M. *Shamanism: Archaic Techniques of Ecstasy.* Princeton: Princeton University Press, 1974.

Fortune, R.F. *Sorcerers of Dobu: The Social Anthropology of the Dobu Islanders of the Western Pacific.* Prospect Heights, IL: Waveland Press, Inc., 1989.

Fuente-Fernández, R. de la, T.J. Ruth, V. Sossi, M. Schulzer, D.B. Calne, A.J. Stoessl. "Expectation and Dopamine Release: Mechanism of the Placebo Effect in Parkinson's Disease." *Science* 293, 1164–66, 2001.

Harrington, A. *The Placebo Effect.* Cambridge, MA: Harvard University Press, 1997.

Harris, W.S., M. Gowda, J.W. Kolb, C.P. Strychacz, J.L. Vacek, P.G. Jones, A. Forker, J.H. O'Keefe, B.D. McCallister. "A Randomized, Controlled Trial of the Effects of Remote, Intercessory Prayer on Outcomes in Patients Admitted to the Coronary Care Unit." *Arch Intern Med.* 159, 2273–78, 1999.

Helm, H.M., J.C. Hays, E.P. Flint, H.G. Koening, D.G. Blazer. "Does Private Religious Activity Prolong Survival? A Six-Year Follow-up Study of 3,851 Older Adults." *J Gerontol A Biol Sci Med Sci* 55, 400–405, 2000.

Hrobjartsson, A., P.C. Gotzsche. "Is the Placebo Powerless? An Analysis of Clinical Trials Comparing Placebo with No Treatment." *New Engl J Med* 344, 1594–1602, 2001.

Hummer, R.A., R.G. Rogers, C.B. Nam, C.G. Ellison. "Religious Involvement and U.S. Adult Mortality." *Demography* 36, 273–85, 1999.

Koenig, H.G., D.B. Larson. "Use of Hospital Services, Religious Attendance, and Religious Affiliation." *South Med J* 91, 925–30, 1998.

Koenig, H.G., J.C. Hays, D.B.Larson, L.K. George, H.J. Cohen, M.E. McCullough, K.G. Meador, D.G. Blazer. "Does Religious Attendance Prolong Survival? A Six-Year Follow-up Study of 3,968 Older Adults." *J Gerontol A Biol Sci Med Sci* 54, 370–76, 1999.

Koenig, H.G. "Psychoneuroimmunology and the Faith Factor." *J Gend Specif* 3, 37–44, 2000.

Krause, N. "Stressors in Highly Valued Roles, Religious Coping, and Mortality." *Psychol Aging.* 13, 242–55, 1998.

Larsen, S. *The Shaman's Doorway: Opening Imagination to Power and Myth.* Barrytown, NY: Station Hill Press, 1988.

Lee, R.B. *The Dobe Ju/'hoansi* (2nd edition). Orlando, FL: Harcourt Brace & Company, 1993.

Levin, J.S. "Religion and Health: Is There an Association, Is It Valid, and Is It Causal?" *Soc Sci Med* 38, 1475–82, 1994.

Levin, J.S., P.L. Schiller. "Is There a Religious Factor in Health?" *J Rel Health* 26, 9–36, 1987.

Loewenthal, K.M., M. Cinnirella, G. Evdoka, P. Murphy. "Faith Conquers All? Beliefs About the Role of Religious Factors in Coping with Depression Among Different Cultural-Religious Groups in the U.K." *Br J Med Psychol* 74, 293–303, 2001.

McCullough, M.E., D.B. Larson. "Religion and Depression: A Review of the Literature." *Twin Res* 2, 126–36, 1999.

McCullough, M.E., W.T. Hoyt, D.B. Larson, H.G. Koenig, C. Thoresen. "Religious Involvement and Morality: A Meta-Analytic Review." *Health Psychol* 19, 211–22, 2000.

Oxman, T.E., D.H. Freeman Jr., E.D. Manheimer. "Lack of Social Participation or Religious Strength and Comfort as Risk Factors for Death After Cardiac Surgery in the Elderly." *Pyschosom Med* 57, 5–15, 1995.

Pargament, K.I., H. Koenig, N. Tarakeshwar, J. Hahn. "Religious Struggle as a Predictor of Mortality Among Medically Ill Elderly Patients." *Arch Intern Med.* 161, 1881–85, 2001.

Rolston III, H. *Genes, Genesis and God.* New York: Cambridge University Press, 1998.

Shapiro, A.K., R. Frick, L. Morris, et al. "Placebo Induced Side Effects." *J Op Psych* 6, 43–46, 1974.

Shapiro, A.K., E. Shapiro. *The Powerful Placebo.* Baltimore and London: The Johns Hopkins University Press, 1997.

Sicher, F., E. Targ, D. Moore II, H.S. Smith. "A Randomized Double-Blind Study of the Effect of Distant Healing in a Population with Advanced AIDS—Report of a Small Scale Study." *WJM* 169, 356–63, 1998.

Sober, E., Wilson, D.S. *Unto Others.* Cambridge, MA: Harvard University Press, 1998.

Stefano, G.B., G.L. Fricchione, B.T. Slingsby, H. Benson. "The Placebo Effect and Relaxation Response: Neural Processes and Their Coupling to Constitutive Nitric Oxide." *Brain Res Rev* 35, 1–19, 2001.

Strawbridge, W.J., R.D. Cohen, S.J. Shema, G.A. Kaplan. "Frequent Attendance at Religious Services and Mortality Over 28 Years." *Am J Public Health* 87, 957–61, 1997.

Strawbridge, W.J., S.J. Shema, R.D. Cohen, G.A. Kaplan. "Religious

Attendance Increases Survival by Improving and Maintaining Good Health Behaviors, Mental Health, and Social Relationships." *Soc Behav Med* 23, 68–74, 2001.

Wilson, D.S., *Darwin's Cathedral: Evolution, Religion and the Nature of Society.* Chicago, IL: The University of Chicago Press, 2002.

Wilson, E.O. *On Human Nature.* Cambridge, MA: Harvard University Press, 1998.

Chapter 9: Religion: From Genes to Memes

Beer, J.M., R.D. Arnold, J.C. Loehlin. "Genetic and Environmental Influences on MMPI Factor Scales: Joint Model Fitting to Twin and Adoption Data." *J Personality Soc Psycho* 74, 818–27, 1998.

Blackmore, S. *The Meme Machine.* New York: Oxford University Press, 1999.

Boomsma, D.I., E.J.C. de Geus, G.C.M. van Baal, J.R. Koopmans. "A Religious Upbringing Reduces the Influence of Genetic Factors on Disinhibition: Evidence for Interaction Between Genotype and Environment on Personality." *Twin Res* 2, 115–25, 1999.

Cavalli-Sforza, L.L., M.W. Feldman. *Cultural Transmission and Evolution: A Quantitative Approach.* Princeton, NJ: Princeton University Press, 1981.

Dawkins, R. *Unweaving the Rainbow: Science, Delusion, and the Appetite for Wonder.* New York: Houghton Mifflin Co., 1998.

Diamond, J. *Guns, Germs, and Steel.* London: Cape, 1997.

Eaves, L., A. Heath, N. Martin, H. Maes, M. Neale, K. Kendler, K. Kirk, L.Corey. "Comparing the Biological and Cultural Inheritance of Personality and Social Attitudes in the Virginia 20,000 Study of Twins and their Relatives." *Twin Res* 2, 62–80, 1999.

Eaves, L.J., N.G. Martin, A.C. Heath. "Religious Affiliation in Twins and Their Parents: Testing a Model of Cultural Inheritance." *Behav Genet* 20, 1–22, 1989.

Kendler, K.S., C.O. Gardner, C.A. Prescott. "Religion, Psychopathology, and Substance Use and Abuse: A Multimeasure, Genetic-Epidemiologic Study." *Am J Psych.* 154, 322–29, 1997.

Kirk, K.M., H.H. Maes, M.C. Neale, A.C. Heath, N.G. Martin, L.J. Eaves. "Frequency of Church Attendance in Australia and the United States: Models of Family Resemblance." *Twin Res* 2, 99–107, 1999.

Lindeman, B. *The Twins Who Found Each Other.* New York: William Morrow and Company, 1969.

Lynch, A. *Thought Contagion: How Belief Spreads Through Society.* New York: Basic Books, 1996.

Maes, H.H., M.C. Neale, N.G. Martin, A.C. Heath, L.J. Eaves. "Religious Attendance and Frequency of Alcohol Use: Same Genes or Same Environments: A Bivariate Extended Twin Kinship Model." *Twin Res* 2, 169–79, 1999.

Martin, N.G., L.J. Eaves, A.C. Heath, R. Jardine, L.M. Feingold, H.J. Eysenck. "Transmission of Social Attitudes." *Proc Nat Acad Sci USA* 83, 4364–68, 1986.

Rose, R. "Genetic and Environmental Variance in Content Dimensions of the MMPI." *J Pers Soc Psychol* 55, 302–11, 1988.

Truett, K.R., L.J. Eaves, J.M. Myer, A.C. Heath, N.G. Martin. "Religion and Education as Mediators of Attitudes: A Multivariate Analysis." *Behav Gene* 22, 43–62, 1992.

Truett, K.R., L.J. Eaves, E.E. Walters, A.C. Heath, J.K. Hewitt, J.M. Meyer, J. Silberg, M.C. Neale, N.G. Martin, K.S. Kendler. "A Model System for Analysis of Family Resemblance in Extended Kinships of Twins." *Behav Gene* 24, 35–49, 1994.

Winter, T., J. Kaprio, R.J. Viken, S. Karvonen, R.J. Rose. "Individual Differences in Adolescent Religiosity in Finland: Familial Effects Are Modified by Sex and Region of Residence." *Twin Res* 2, 108–14, 1999.

Chapter 10: The DNA of the Jews

Bamshad, M.J., W.S. Watkins, M.E. Dixon, L.B. Jorde, B.B. Rao, J.M. Naidu, B.V. Prasad, A. Rasanayagam, M.F. Hammer. "Female Gene Flow Stratifies Hindu Castes." *Nature* 395, 651–52, 1998.

Bamshad, M., T. Kivisild, W.S. Watkins, M.E. Dixon, C.E. Ricker, B.B. Rao, J.M. Naidu, B.V. Prasad, P.G. Reddy, A. Rasanayagam, S.S. Papiha, R. Villems, A.J. Redd, M.F. Hammer, S.V. Nguyen, M.L. Carroll, M.A. Batzer, L.B. Jorde. "Genetic Evidence on the Origins of Indian Caste Populations." *Genome Res* 11, 994–1004, 2001.

Behar, D.M., M.G. Thomas, K. Skorecki, M.F. Hammer, E. Bulygina, D. Rosengarten, A.L. Jones, K. Held, V. Moses, D. Goldstein, N. Bradman, M.E. Weale. "Multiple Origins of Ashkenazi Levites: Y Chromosome Evidence for Both Near Eastern and European Ancestries." *Am J Hum Genet* 73, 768–79, 2003.

Behar, D.M., M.F. Hammer, D. Garrigan, R. Villems, B. Bonne-Tamir, M. Richards, D. Gurwitz, D. Rosengarten, M. Kaplan, S.D. Pergola, L. Quintana-Murci, K. Skorecki. "MtDNA Evidence for a Genetic Bottleneck in the Early History of the Ashkenazi Jewish Population." *Eur J Hum Genet*, 2004, in press.

Behar, D.M., D. Garrigan, M.E. Kaplan, Z. Mobasher, D. Rosengarten, T.M. Karafet, L. Quintana-Murci, H. Ostrer, K. Skorecki, M.F. Hammer. "Contrasting Patterns of Y Chromosome Variation in Ashkenazi Jewish and Host Non-Jewish European Populations." *Hum Genet* 114, 354–65, 2004.

Foster, E.A., M.A. Jobling, P.G. Taylor, P. Donnelly, P. de Knijff, R. Mieremet, T. Zerjal, C. Tyler-Smith. "Jefferson Fathered Slave's Last Child." *Nature* 396, 27–28, 1998.

Grayzel, S. *A History of the Jews*. New York: Penguin, 1947.

Hammer, M.F. "A Recent Common Ancestry for Human Y Chromosomes." *Nature* 378, 376–78, 1995.

Hammer, M.F., A.B. Spurdle, T. Karafet, M.R. Bonner, E.T. Wood, A. Novelletto, P. Malaspina, R.J. Mitchell, S. Horai, T. Jenkins, S.L. Zegura. "The Geographic Distribution of Human Y Chromosome Variation." *Genetics* 145, 787–805, 1997.

Hammer, M.F., A.J. Redd, E.T. Wood, M.R. Bonner, H. Jarjanazi, T. Karafet, S. Santachiara-Benerecetti, A. Oppenheim, M.A. Jobling, T. Jenkins, H. Ostrer, B. Bonne-Tamir. "Jewish and Middle Eastern Non-Jewish Populations Share a Common Pool of Y-Chromosome Biallelic Haplotypes." *Proc Natl Acad Sci USA* 97, 6769–74, 2000.

Hammer, M.F., F. Blackmer, D. Garrigan, M.W. Nachman, J.A. Wilder. "Human Population Structure and Its Effects on Sampling Y Chromosome Sequence Variation." *Genetics* 164, 1495–509, 2003.

Jobling, M.A., C. Tyler-Smith. "Fathers and Sons: The Y Chromosome and Human Evolution." *Trends Genet* 11, 449–56, 1995.

Kobyliansky, E., S. Micle, M. Goldschmidt-Nathan, B. Arensburg, H. Nathan. "Jewish Populations of the World: Genetic Likeness and Differences." *Ann Hum Biol* 9, 1–34, 1982.

Lucotte, G., P. Smets, J. Ruffie. "Y-Chromosome-Specific Haplotype Diversity in Ashkenazic and Sephardic Jews." *Hum Biol* 65, 835–40, 1993.

Lucotte, G., F. David, S. Berriche. "Haplotype VIII of the Y Chromosome Is the Ancestral Haplotype in Jews." *Hum Biol* 68, 467–71, 1996.

Lucotte, G., P. Smets. "Origins of Falasha Jews Studied by Haplotypes of the Y Chromosome." *Hum Biol* 71, 989–93, 1999.

Lucotte, G., G. Mercier. "Y-Chromosome DNA Haplotypes in Jews: Comparisons with Lebanese and Palestinians." *Genet Test* 7, 67–71, 2003.

Mitchell, R.J., M.F. Hammer. "Human Evolution and the Y Chromosome." *Curr Opin Genet Dev* 6, 737–42, 1996.

Parfitt, T. *Journey to the Vanished City*. London: Phoenix, 1997.

Parfitt, T. "Constructing Black Jews: Genetic Tests and the Lemba—the 'Black Jews' of South Africa." *Dev World Bioeth* 3, 112–18, 2003.

Risch, N., D. de Leon, L. Ozelius, P. Kramer, L. Almasy, B. Singer, S. Fahn, X. Breakefield, S. Bressman. "Genetic Analysis of Idiopathic Torsion Dystonia in Ashkenazi Jews and Their Recent Descent from a Small Founder Population." *Nat Genet* 9, 2, 152–59, 1995.

Ritte, U., E. Neufeld, M. Broit, D. Shavit, U. Motro. "The Differences Among Jewish Communities—Maternal and Paternal Contributions." *J Mol Evol* 37, 435–40, 1993.

Santachiara-Benerecetti, A.S., O. Semino, G. Passarino, A. Torroni, R. Brdicka, M. Fellous, G. Modiano. "The Common, Near-Eastern Origin of Ashkenazi and Sephardi Jews Supported by Y-Chromosome Similarity." *Ann Hum Genet* 57, 55–4, 1993.

Skorecki, K., S. Selig, S. Blazer, R. Bradman, N. Bradman, P.J. Waburton, M. Ismajlowicz, M.F. Hammer. "Y Chromosomes of Jewish Priests." *Nature* 385, 32, 1997.

Spurdle, A.B., T. Jenkins. "The Y Chromosome as a Tool for Studying Human Evolution." *Curr Opin Genet Dev* 2, 487–91, 1992.

Spurdle, A., T. Jenkins. "Y Chromosome Probe p49a Detects Complex PvuII Haplotypes and Many New TaqI Haplotypes in Southern African Populations." *Am J Hum Genet* 50, 107–25, 1992.

Spurdle, A.B., M.F. Hammer, T. Jenkins. "The Y Alu Polymorphism in Southern African Populations and Its Relationship to Other Y-Specific Polymorphisms." *Am J Hum Genet* 54, 319–30, 1994.

Spurdle, A.B., T. Jenkins. "The Origins of the Lemba 'Black Jews' of Southern Africa: Evidence from p12F2 and Other Y-Chromosome Markers." *Am J Hum Genet* 59, 1126–33, 1996.

Sykes, B., C. Irven. "Surnames and the Y Chromosome." *Am J Hum Genet* 66, 1417–19, 2000.

Thomas, M.G., K. Skorecki, H. Ben-Ami, T. Parfitt, N. Bradman, D.B.

Goldstein. "Origins of Old Testament Priests." *Nature* 394, 138–40, 1998.

Thomas, M.G., T. Parfitt, D.A. Weiss, K. Skorecki, J.F. Wilson, M. le Roux, N. Bradman, D.B. Goldstein. "Y Chromosomes Traveling South: The Cohen Modal Haplotype and the Origins of the Lemba—the 'Black Jews' of Southern Africa." *Am J Hum Genet* 66, 674–86, 2000.

Tikochinski, Y., U. Ritte, S.R. Gross, E.M. Prager, A.C. Wilson. "mtDNA Polymorphism in Two Communities of Jews." *Am J Hum Genet* 48, 129–36, 1991.

Chapter II: God Is Alive

Barbour, I.G. *Religion and Science.* San Francisco: HarperCollins, 1997.

Boyer, P. *The Naturalness of Religious Ideas: A Cognitive Theory of Religion.* Berkeley and Los Angeles: University of California Press, 1994.

Brown, D.E. *Human Universals.* New York: McGraw-Hill, 1991.

Davies, P. *The Mind of God: The Scientific Basis for a Rational World.* New York: Touchstone, 1993.

Dawkins, R. *The Selfish Gene.* New York: Oxford University Press, 1976.

Gibson, A., D. Simpson. *Prehistoric Ritual and Religion.* Gloucestershire, UK: Sutton Publishing Ltd., 1998.

Glynn, P. *God—The Evidence: The Reconciliation of Faith and Reason in a Postsecular World.* Rocklin, CA: Prima Publishing, 1999.

Gould, S.J. *Rocks of Ages: Science and Religion and the Fullness of Life.* New York: The Ballantine Publishing Group, 1999.

Morris, B. *Anthropological Studies of Religion: An Introductory Text.* Cambridge, UK: The Cambridge University Press, 1987.

Pinker, S. *How the Mind Works.* New York: W.W. Norton & Company, Inc., 1997.

Polkinghorne, J. *Belief in God in an Age of Science.* New Haven: Yale University Press, 1998.

Rappaport, R.A. *Ritual and Religion in the Making of Humanity.* Cambridge, UK: Cambridge University Press, 1999.

Russell, B. *Religion and Science.* New York: Oxford University Press, 1997.

Smith, H. *Why Religion Matters: The Fate of the Human Spirit in an Age of Disbelief.* San Francisco: HarperCollins, 2001.

Templeton, J.M., Herrmann, R.L. *The God Who Would Be Known.* Radnor, PA: Templeton Foundation Press, 1998.

Wilson, E.O. *Consilience: The Unity of Knowledge.* New York: Alfred A. Knopf, Inc., 1998.

Wilson, E.O. *On Human Nature.* Cambridge: Harvard University Press, 1998.

Yu-Lan, F. *A Short History of Chinese Philosophy.* New York: The Free Press, 1966.

Index

N

O

P

R

S

T